CONCILIUM

Religion in the Eighties

D1427626

CONCILIUM

Concilium 166 (6/1983): Project 'X'

COSMOLOGY
AND
THEOLOGY

Edited by

David Tracy

and

Nicholas Lash

English Language Editor
Marcus Lefébure

T. & T. CLARK LTD
Edinburgh

THE SEABURY PRESS
New York

June 1983
T. & T. Clark Ltd, 36 George Street, Edinburgh EH2 2LQ
ISBN: 0 567 30046 3

The Seabury Press, 815 Second Avenue, New York, NY 10017
ISBN: 0 8164 2446 2

Library of Congress Catalog Card No.: 82 062759

Printed in Scotland by William Blackwood & Sons Ltd, Edinburgh

Concilium: Monthly except July and August.
Subscriptions 1983: UK and Rest of the World £27·00, postage and handling included;
USA and Canada, all applications for subscriptions and enquiries about *Concilium*
should be addressed to The Seabury Press, 815 Second Avenue, New York, NY 10017,
USA.

CONTENTS

Part III
Concluding Editorial Reflections

Editorial

THE 'POSITIVISM' which supposed that only the physical sciences (and disciplines which conform to their procedures) could furnish us with *knowledge* of the world is on the wane, although it remains influential in the popular imagination. What is taking its place, and what are the implications for theology of the recognition that scientific understanding is, as are all forms of human understanding, shaped by need, hope, time and circumstance?

Since the eighteenth century theologians have tended, partly as a result of the 'anthropologising' and 'individualising' of theological concern, to concentrate on doctrines of redemption to the neglect of doctrines of creation. But what should be the relationship between these two aspects of Christian belief? In preparing this issue, it became clear to us that questions of 'cosmology' concern not only the origin and 'natural structure' of the world but also its destiny. When 'creation' and 'redemption' are too sharply separated, reflection on the latter neglects the implications of the fact that the destiny of human beings, of 'animals who hope', is inextricably bound up with the destiny of a cosmos for which we share responsibility.

A further difficulty: scientific and theological understanding may not be necessarily conflictual, but neither may they be conflated into the grammar of a single discourse. Attempts to do so usually seek to 'fit' religious symbols into some larger pattern of scientific discourse. The result has often been to understate the tragic. Especially in the nineteenth century, religious world views derived from the supposedly 'value free' descriptions of science tended to be unwarrantedly 'optimistic' in character.

'Cosmology' may mean many things. The term can refer to theological accounts of the world as God's creation; or to philosophical reflection on the categories of space and time; or to observational and theoretical study of the structure and evolution of the physical universe; or, finally, to 'world views': unified imaginative perceptions of how the world seems and where we stand in it.

We therefore decided to place the initial emphasis on the *history* of the relationships between theological and cosmological doctrines. Without an attempt to 'retrieve' the history, the current state of the problem cannot be appropriately indicated.

J. Collins, in the first of a group of articles which concentrates on the historical issues, emphasises the enduring validity of the insight that the question of human salvation cannot be divorced from our understanding of and relationship to the world around us. A similar consideration emerges from *H. Chadwick's* discussion of the relationships between freedom and necessity in early Christian thought about God. *O. Pedersen* highlights the role played by theology in helping to liberate human beings from the fixed confines of the ancient cosmos only to release them into those infinite spaces in which God was not found. Once the connections between God and space had been severed the category of the *temporal* came increasingly to dominate scientific and theological discussion (see *G. Altner*). To raise the question of time is to raise the question of 'end-time'. *Bp Tshishiku*, therefore, discusses some developments in eschatology.

J. Buchanan, in the first of a group of articles dealing with particular topics, argues for the indispensability of the symbol of 'creation' for holding in tension our sense of participation in and distinctness from the natural order of the world. *H. Brück* summarises developments in scientific cosmology, and insists that scientific exploration of the 'beginning' of the universe is not to be confused with consideration of its 'creation' by God. When cosmology and theology conflict, suggests *M. Hesse*, they do so because neither discourse has sufficiently acknowledged its status as symbolic construction of reality. This thesis is illustrated by *L. Gilkey's* reflections on disputes between 'creationism' and 'evolution'. The length of his article is due to the fact that, when another contributor had been unable to accept our invitation, Gilkey kindly agreed to locate these disputes in the broader context of the relationship between scientific and religious 'world views'. There follows an assessment of Teilhard de Chardin's contribution to our theme (see *W. Warthling*). Finally, because some Western scientists, in becoming conscious of the need to attend to the 'spiritual', have begun to look to the East rather than to Western religion, *U. King* has provided a Bulletin on 'Modern Cosmology and Eastern Thought'.

How is the universal (cosmic?) significance of the Incarnation to be understood in a world for which no single, all-embracing, 'scientific' cosmological narrative can be constructed? What forms of hope in God the Creator are appropriate in a world in which ideologies of 'progress' have perished? These are just two of the questions which we had originally hoped to cover, and on some of which we have touched in our 'Concluding Editorial Reflections'.

DAVID TRACY
NICHOLAS LASH

PART I

Historical

John Collins

New Testament Cosmology

'FOR THE creation waits with eager longing for the revelation of the sons of God; for the creation was subjected to futility, not of its own will but by the will of him who subjected it in hope, because the creation itself will be set free from its bondage to decay and obtain the glorious liberty of the children of God' (Rom. 8:19-21).

This statement by Paul is remarkable for its affirmation that human destiny is bound up with the physical creation. It is all the more remarkable in view of the widely held opinion that the Old Testament attaches little importance to cosmology and locates revelation in history. To be sure, the latter opinion is exaggerated. The theology of the Psalms and Wisdom books is based on creation rather than on history and cosmic imagery abounds in the prophets. Yet it is true that cosmology plays a more prominent part in the New Testament than in the Hebrew Bible. In fact, the Jewish traditions inherited by early Christianity had undergone major developments in the Hellenistic age. These developments are most readily apparent in two bodies of literature which were especially influential on the New Testament: the Wisdom books and the apocalypses.

1. WISDOM

The Wisdom tradition in the Old Testament had always looked to creation rather than the history of Israel for the foundation of its theology. According to Proverbs 8, Wisdom was set up before the beginning of the earth and was there when God established the heaven. Wisdom is embedded in the universe and provides a link between God and humanity. The wise man who finds wisdom thereby makes contact with God. According to Sirach (Ecclesiasticus) 24, this Wisdom, which has made the circuit of heaven and walked in the depths of the sea, has made its special dwelling in Israel, specifically in the book of the law of Moses. Yet it is also true that 'in every people and nation I have gotten a possession' (Ecclus. 24:6). Both Proverbs and Sirach attempt to relate the wisdom which leads to righteousness in human actions to a cosmic principle, present in creation. Neither book attempts to provide a metaphysic, to explain how this could be, but is content to personify Wisdom as a quasi-mythological figure.

The transformation of the Wisdom tradition in the Hellenistic age is seen most clearly in the Wisdom of Solomon, which was written in Greek about the time of Christ, most probably in Alexandria. There Wisdom takes on some of the character of the Stoic *Logos*. The *Logos* or *Pneuma* (Spirit) was conceived as a very fine fiery substance which is transfused through passive matter and organises it. The Logos is at once God, Nature and Reason. Human reason was thought to share literally and physically in the Divine Nature. The ideal of the Stoic wise man was to put himself in harmony with this cosmic principle by living according to nature or reason. Naturally, this notion of the Logos undergoes some modification in a Jewish context but its influence on the Wisdom of Solomon is undeniable. Wisdom is now presented in quasi-physical terms: 'For wisdom is more mobile than any motion; because of her pureness she pervades and penetrates all things' (7:24). 'She reaches mightily from one end of the earth to the other, and she orders all things well' (8:1). Wisdom is evidently identical with the 'Spirit of the Lord' which fills the world and holds all things together (1:7).

This cosmic force of Wisdom, which is synonymous with the Word or Logos (9:1, 2) influences human destiny in a number of ways. On the individual level: 'generation by generation passing into holy souls she makes them friends of God and prophets' (7:27). It is Wisdom that makes the soul immortal. Moreover Wisdom is the agent of salvation in history: 'A holy people and a blameless race wisdom delivered from a nation of oppressors . . .' (10:15). The experience of Israel and its enemies is expressed as an experience of nature rather than a direct encounter with God. The plagues of Egypt and the crossing of the sea reveal that 'nature fights for the righteous' (15:17) and that 'creation, ministering to thee its maker, strains itself against the unrighteous for punishment and slackens for beneficence on behalf of those that trust in you [God]' (16:24). Even miracles do not disrupt the order of nature, but re-arrange it: 'For the elements changed places with one another as on a harp the notes vary the nature of the rhythm while each note remains the same' (19:18).

In a world so constituted, the key to salvation is understanding of the workings of the cosmos. Solomon's glory is that God 'gave me unerring knowledge of what exists, to know the structure of the world and the activity of the elements; the beginning and end and middle of times' (7:17-18). The wicked are misled because they do not know the mysteries of God (2:22). Conversely, those who trust in God will understand truth (3:9) and so their hope is full of immortality.

2. THE APOCALYPSES

The Jewish apocalypses have no unifying cosmic figure like Wisdom, but they share the conviction that salvation depends on a proper understanding of the workings of the cosmos. The genre apocalypse first appears in Judaism in the Hellenistic age. The earliest examples are found in the Books of Enoch, some of which show heavy Babylonian influence. This point is of interest because Babylon was famous for the development of astronomy and astrology and the conviction that historical events were related to the movements of the stars. Interest in the astral world is prominent in the earliest Enoch books.

While the apocalypses are related to biblical prophecy in their interest in visions and focus on eschatology, their idea of revelation is quite different. The apocalyptic seer, Enoch or Daniel, is given mysterious visions, which must be explained to him by an angel. The emphasis is on understanding, not directly on obedience, as in the prophetic texts. Significantly the major apocalyptic seers, Enoch, Daniel, Ezra, Baruch, were all sages rather than prophets. The heroes of the Book of Daniel were *maskilim* or wise teachers. Apocalyptic revelation, in short, is a kind of wisdom. At the same time it differs from Proverbs and the Wisdom of Solomon in so far as it is revealed wisdom which requires angelic intervention.

The Jewish apocalypses may be divided into two main types. The first has an historical orientation and is exemplified by the Book of Daniel or 4 Ezra. The second involves a heavenly journey. Examples are found in 1 Enoch, 2 Enoch and 3 Baruch. The 'historical' apocalypses are sparing in cosmological detail but nevertheless imply a definite structure to the universe. To begin, there are two levels of reality, one occupied by human beings and the other by angels and supernatural forces. By visions, and the angel's explanation, the visionary is allowed to see the course of history from the angelic level. From this level it is apparent that the course of history is pre-determined. Often it is divided into set periods, such as the famous four kingdoms of Daniel. At a pre-determined time there will be a great crisis, followed by resurrection and the judgment of the dead. In the later apocalypses of the first century CE, such as 4 Ezra, the crisis involves the end of this world and a new creation. Even in Daniel, which gives only minimal details about the final transformation, it is apparent that the present world-order will come to an end and be replaced by a new one.

The apocalypses which recount heavenly journeys contain far more cosmological detail than Daniel or 4 Ezra. Enoch is guided by angels to the ends of the earth. He is shown the cornerstone of the earth and the four winds which support the firmament of heaven (1 Enoch 18). He is also shown the prison for the fallen angels, and the resting-places of the spirits of the souls of the dead (chap. 22). It is important that the places of judgment are already prepared and are built in to the structure of the universe. In later apocalypses, such as 2 Enoch, the heavens are displayed in an orderly, numbered sequence, usually seven, in a manner analogous to the division of history into set periods. All these apocalypses envisage a sharp contrast between the order of the heavens and the disorder on earth. A few (e.g., 2 Enoch) also envisage that the present world will pass away.

The cosmology of the apocalypses bears directly on human salvation. In an historical apocalypse like Daniel it is essential that the duration of history is pre-determined. The troubles of the present will not last forever. An important feature of these apocalypses is the insistence that the time of the end is near. It is also essential that, in the phrase of 4 Ezra, 'this present world is not the end' (4 Ezra 7:112). Whether the other world is conceived primarily in spatial terms, as in the heavenly journeys, or in temporal terms as the world to come, it provides the alternative to the present order which makes salvation possible. In fact, both the spatial and the temporal dimensions are represented in all the apocalypses.

As in Wisdom, the key to salvation is understanding. The revelation of the heavenly mysteries by Enoch or Daniel provides the perspective which enables the wise to lose their lives, if necessary, in the present. If there is a judgment beyond death, then this life is not all important. Other values take precedence over self-preservation. This perspective involved a profound shift from the values of the Hebrew Bible, where salvation was often conceived in terms of long life in the land and plentiful descendants. For the ancient Israelite, there was indeed a world beyond this one, in the heavenly council of Yahweh and in the shadowy netherworld of Sheol. However, there was no possibility for a vital and meaningful human life beyond death. The crucial difference is that in the Hellenistic age humanity can hope for admittance to the heavenly world and, in the words of Enoch, become companions to the host of heaven (1 Enoch 104:6).

3. THE NEW TESTAMENT

The world view of the apocalypses was widespread in Judaism at the turn of the era, even in documents which were not apocalypses, as can be seen from the Dead Sea Scrolls. It is quite central to the New Testament. Only one document in the New

Testament is an apocalypse: the Revelation to John, which ends with a vision of a new heaven and a new earth. The importance of Revelation should not be underestimated but it is scarcely at the centre of the New Testament. The apocalyptic world view, however, is not confined to this one book. It underlies the most central of all Christian beliefs, the resurrection.

In his great discourse on the resurrection in 1 Corinthians 15, Paul ties the resurrection of Jesus firmly to the apocalyptic notion of a general resurrection. 'If there is no resurrection of the dead, then Christ has not been raised' (1 Cor. 15:13). Neither an empty tomb not visions could carry conviction if one denied in principle that resurrection was possible. Rather, Christ is viewed as 'the first fruits of those who have fallen asleep'. The resurrection of Jesus is credible because of the apocalyptic world view, but it also confirms the more general expectation. What is at issue is the reality of the world beyond this one: 'If for this life only we have hoped in Christ, we are of all men most to be pitied.'

The hope for a life beyond this one presupposes that 'this present world is not the end'. Paul was convinced that 'the form of this world is passing away': 'I mean, brethren, the appointed time has grown very short; from now on let those who have wives live as though they had none . . . and those who deal with the world as though they had not dealings with it' (1 Cor. 7:29-32). The conviction that whis world is passing away is also evident in the gospels, where we read that 'the sun will be darkened, and the moon will not give its light and the stars will be falling from heaven and the powers in heaven will be shaken. And then they will see the Son of Man coming in clouds with great power and glory. . . . Truly, I say to you, this generation will not pass away before all these things take place. Heaven and earth will pass away, but my words will not pass away' (Mark 13:24-30). The most elaborate descriptions of the passing of this world are found, of course, in Revelation.

The New Testament is quite reticent about the world beyond. Paul refers enigmatically to a rapture to 'the third heaven' to Paradise, but gives no account of what was seen there (2 Cor. 12:2-3). He is also confident that 'if the earthly tent we live in is destroyed, we have a building from God, a house not made with hands, eternal in the heavens' (2 Cor. 5:1). Only Revelation attempts to elaborate on the new creation (Rev. 21, 22). Yet, the reality of the world to come, and of the heavenly principalities and powers, is fundamental to the New Testament throughout. In the second century the cosmological presupposition of Christianity receive more explicit attention with the proliferation of Christian apocalypses.

Christianity differed from the Jewish apocalypses by the central role accorded to Jesus Christ. In the gospels this role is expressed primarily in the mythological terms. Jesus is the Son of Man coming on the clouds of heaven, like the enigmatic 'one like a son of man' in Daniel. However, the significance of Jesus for Christianity was not only that he would be saviour and judge. He was also the model to be imitated. So Paul depicts Jesus as the paradigm for a new humanity, a new Adam (Romans 5). Increasingly, in the later books of the New Testament, he is given cosmic significance. Christ is at once the revealer of the apocalyptic mysteries and their man subject: 'For he has made known to us in all wisdom and insight the mystery of his will, according to his purpose which he set forth in Christ, as a plan for the fullness of time, to unite all things in him, things in heaven and things on earth' (Ephesians 1:9-10). Christ here becomes 'that which holds all things together', like Widsom in the Wisdom of Solomon. This idea is elaborated in Colossians: 'He is the image of the invisible God, the first-born of all creation; for in him all things were created, in heaven and on earth, visible and invisible, whether thrones or dominations, or principalities or authorities—all things were created through him and for him. He is before all things, and in him all things hold together' (Col. 1:15-17). Christ is a cosmological principle who mediates between God and the world very much in the manner of Wisdom. The link with the apocalyptic view is apparent in Colossians 3: 'for you have died and your life is hid with Christ in God'. The idea that the ultimate structure of the universe is a mystery and involves a hidden dimension of reality is a point of contact between the apocalyptic mysteries and the 'mysteries of God' of the Wisdom of Solomon.

The most obvious fusion of cosmology and Christology in the New Testament is found in the prologue of the Gospel of John. There the links with apocalyptic thought are less apparent, but the influence of the Wisdom tradition is clear. Christ is the Logos which was in the beginning with God. John goes beyond the Wisdom tradition in the bold equation 'and the Logos was God'. All things were made through him and he was in the world, although the world knew him not. Finally the Logos takes flesh in the person of Jesus, as Wisdom in Sirach 24 had made its dwelling in Zion and the Jewish law. From this perspective, Jesus of Nazareth is the supreme expression of a cosmic principle, which was present in the world long before Jesus was born. Christian life, then, is an attunement to this cosmic principle, just as Stoic wisdom was, although the specific content of Christian wisdom is quite different from that of the Stoics.

4. CONCLUSION

In the preceding pages we have attempted to sketch, all too rapidly, some of the basic factors in the New Testament world view. The thrust of our argument has been that in two major areas of Hellenistic Judaism salvation was bound up with the understanding of the cosmos. The New Testament inherited this conviction, and it shaped the development of Christian eschatology and of Christology. In conclusion we must ask what is the abiding significance of New Testament cosmology for Christian faith?

Rudolf Bultmann, the most influential New Testament scholar of this century, viewed the cosmology as a husk from which the existential message of individual salvation must be extracted. More recent scholarship has tended to affirm that the cosmology is an integral part of the New Testament message, and cannot be dismissed as of secondary importance. Bultmann's position had some obvious merits. The cosmology of the New Testament is untenable today. It is apparent from the Jewish apocalypses that the predictions of the end and the visious of the heavens are imaginative constructs. They do not in fact give cosmological information but provide a way of imagining the world which gives support to a certain value system. The same must be said of Paul and the Synoptics. Again the notion of the Logos in Christianity is primarily a way of affirming the ultimacy of the wisdom of Jesus. It would require substantial revision to bring it into line with modern cosmology. However, Bultmann's critics have also made a valid point. *Human salvation cannot be divorced from our understanding of the world around us.* The creation, too, is groaning in travail. Our total view of the world will no doubt be always symbolic in character and be a reflection of the values we wish to affirm. It is important, however, that we find a way to integrate our human values with some cosmological understanding if our theology is to represent more than a fragment of our experience.

Henry Chadwick

Freedom and Necessity in Early Christian Thought About God

THE ATTRIBUTION to a divine Mind of the origination and subsequent maintaining of the cosmos is not an idea first launched upon an astonished world by the early Christians. Among the ancient pre-Christian writers of the Greco-Roman world many had much to say about both creation and providence, and what they had to say is in some respects strikingly similar to themes the early Christians were glad to make their own. In other respects, however, the ancient pre-Christian writers need to be heard without any Christian presuppositions being read back into the texts. For the Christians virtually took for granted that this world is brought into being by a benevolent and powerful Creator whose care for humanity, long exercised in a particular, selective way through the giving of the Torah and the call of Israel, has now become manifest as a general care for the entire race. He wills all men to be saved. There is none for whom Christ did not die. Therefore the early Christian notion of both creation and providence is determined by their central affirmation of redemption through the Word made flesh, sharing in our suffering, making a perfect offering of himself to the Father, calling into being a community from all nations, mediating the benefits of his redemption to the individual through grace covenanted and applied through his sacraments of baptism and Eucharist. So in the prologue to St John's gospel, the Word that is made flesh is also the light who lightens every man coming into the world; he is the principle of rationality and order. The particularity of the incarnation is the route through which we may know the universality of divine providence. And conversely, belief in the origination and care of the cosmos as the continuing work of God makes it possible for the understanding to grasp that unique and special care manifest in the incarnation.

No such ideas played any part in pre-Christian debates about the making of the world or of the soul or about the interest of the cosmos's designer in the survival of the species inhabiting the earth. Nevertheless the discussions in Plato, Aristotle, the Stoics, and Epicurus contributed much to determining the form or framework of the Christian understanding of these matters.

The gods have radical exemption from any sort of responsibility for the origination and maintenance of this world in the thought of *Epicurus*. For him the gods exist, but their life is one of tranquillity, of agreeable cultivated conversation untroubled by all the fuss and bother of the busy world over the Garden wall. The natural environment has come about in consequence of a random series of collisions among the atoms, and it is a mistake to start looking for evidence of a designing mind. The *Stoics*, on the other hand, vigorously defended providence. One cannot achieve happiness without confident belief that all is for the best. Zeno and Chrysippus saw evidence of providence in the ordered design of things, in the interlocking chain of causality, in the vitalism of the world's power and beauty. To live according to reason is simultaneously to live according to nature; that is, to accept whatever happens as determining the framework within which one is to do one's duty. Affirming providence is a grateful and also a responsible way of evaluating an environment which one cannot possibly conquer and therefore can only join. The Stoics asserted (without convincing their many critics) that they believed in free will. But this freedom of choice is not in principle different from a choice made by a dog when faced by a fork in the pathway. At times of high crisis, when one is being badly pushed around by hostile forces, the Stoic sage seeks to assert the freedom of his unconquerable soul, and even allows himself the option of suicide if that is wholly and exclusively an act intended to affirm one's ultimate right to remain free in a society seeking to impose unacceptable restraints and conditions. The Stoics did not think of Zeus as having made positive decisions to create the world. The divine 'will' is the unalterable course of nature and the historical process. Fate and Providence are indistinguishable.

Plato's *Timaeus* provided the ancient world with its principal handbook on cosmogony. It was the only dialogue of Plato to pass into Latin, in Cicero's translation of its first half, before the twelfth century. The dialogue is a kind of prose-poem but with a scientific, descriptive intention. The supreme Father is the first cause whose creating act is the consequence of a flowing out of his own goodness. He is 'free of jealousy' (*Tim.* 29e), and therefore 'desired that all things be good, and nothing bad so far as this was attainable'. 'Out of disorder he brought order', by putting *nous* in soul and soul into body.

A world created by pure goodness and itself only good might be expected to need next to no maintenance. The presence of evil in the world is therefore an acute difficulty for a Platonic metaphysic, and the answer to the difficulty takes the form of an elaborate doctrine of the hierarchy of being. The supreme apex of this hierarchy is 'being or beyond being'; all subsequently derived beings have a metaphysical and a moral inferiority—they enjoy less being and less goodness, until we descend to the unformed sludge of unshaped matter whose 'evil' is, strictly, 'non-being'. In the *Republic* Plato vigorously denies that the good Creator can be responsible for evil which, here, comes from a misuse of free choice. In the *Phaedrus* the soul's fall or loss of wings is the consequence of neglect.

The Platonic doctrine of God leaves one with a tension: on the one hand, the cosmos is the natural outflow of his goodness, yet on the other hand he is wholly self-sufficient, in need of nothing. God in Platonism is not vulnerable to any element of chance or necessity. The ills that flesh is heir to do not touch God.

On freedom and necessity one of the most influential tracts was written by *Plotinus* in the third century A.D. (*Enneads* vi, 8). The tract illustrates the problems as a Platonist wrestles with the reconciliation of Plato's authoritative texts. His fundamental thesis is that the All-Highest is good not in the sense that goodness is an accidental quality of his character, but as the very Idea of the Good in such sense that with him to be and to be good are one and the same thing. In the case of derived entities lower down in the hierarchy of being, however, to be is one thing, to be good is another.

For Plotinus it is self-evident that the supreme first cause is not restricted by matter, space, and time. The pressure of physical need can be wholly discounted. Nous is wholly independent when not involved in the material world; and Psyche can become free as it cuts itself off from matter and ascends to Nous. The source of all goodness is not subject to necessity because it cannot move down to the worse. For where there is most good, there is also most freedom (vi, 8, 7). Therefore in the One, at the apex of the hierarchy, essence and will are identical (8, 13).

This last proposition is twice echoed by *Augustine* in the *Confessions* (xi, 10:12; xii, 15:18). Augustine is anxious to avoid the horns of the dilemma that (*a*) if creation issues from the sole goodness of God by a spontaneous outflow, then its existence seems not to result from a decision of the divine but is, in some real sense, an inevitable, almost physical emanation or projection; while (*b*) if the creation is the consequence of an omnipotent decision of transcendent power by a wholly self-sufficient First Cause which does not in any sense whatever need the created order (e.g., to be the object of his love), then the decision comes to look arbitrary, an unexplained and unexplicable suddenness.

The difficulty no doubt arises in large part from the anthropomorphic images which tend to lie behind the debate—an anthropomorphism of which Plotinus was very conscious and against which he utters express warnings (vi, 8:13). 'Emanation' is a physical image from a language-framework associated with (e.g.) the diffusion of light or the growth of a plant. An arbitrary decision in a human situation tends to be one where no reasons either are or can be given. Nevertheless we expect an arbitrator in a dispute to exercise his judgment after taking account of all the facts, and though the final decision depends on the arbitrator's judgment (for judgment is precisely what he is appointed to exercise), it is normally not absurd to ask if his judgment was also reasonable in the circumstances. And in a dispute he not only has to ask himself what is just and right, but also must consider the consequences of his decision and the likelihood of it being accepted by the disputing parties. In short the more an arbitrator will be seen to be offering good judgments, the less arbitrary his decisions seem to be.

The language of divine decision was congenial enough to Augustine when he was talking about the election of grace. He did not think it useful or sensible to ask if any act of will could be exhaustively described in terms of its causation. To ask why Adam's will turned away from immutable Good to mutable goods is to ask an unanswerable question (*De libero arbitrio* ii, 20:54). The will seems to Augustine like gravity, a weight that pulls one in a certain direction, so that one is 'carried along'. (Hence his famous 'feror' in the *Confessions* xiii, 9, 10.) His version of the gospel saying 'where your treasure is, there will your heart be also' is in effect 'my love is the weight which moves me to act'. This sounds uncommonly close to determinism; and it appears repeatedly in Augustine's doctrines of both God and man that he does not feel at ease with the disjunction of nature and will. The Arian Eunomius, for example, is severely rebuked for working with a doctrine of God in which nature is one thing, will another (*De Trinitate* xv, 20:38). In the doctrine of man, the pull on the soul may take it either up or down, to heaven or to hell. The pull of the latter comes from 'bad habit' (*mala consuetudo*), the force of which Augustine sees as in practice unbreakable without the intervention of divine grace. For habit is second nature, so that, for much of the time at least, choice is a merely theoretical possibility and capacity. I *can* do many things, but their achievement (Augustine saw) depends on my *wanting* to do them. That is to say that the driving force producing action is a predominant weight or mass of desire to bring about the acts concerned; action is the product of continuing character, itself formed by the pattern of previous actions (whether on my part or on that of my ancestors). Yet Augustine will never allow the strictly determinist account of human nature to take over. When he has conceded all to the pressures of emotion and habit which we bring into situations of moral decision, he still insists that we remain responsible and retain the power of choice, and that there is that in the choice of will which somehow escapes the explanations of causality.

With this anthropology in the background it becomes easier to understand why Augustine found so much that was congenial to him in Plotinus' proposition that in God the substance and the will are indistinguishable. God is Love, and the strong doctrine that Amor is the very essence of God's being carries the implication that his decision to create cannot but spring from the goodness of his will. But neither in God nor in man is the will some sort of external rudder steering the ship. Plotinus (vi, 8:13) insists that in God essence and act stand in the most intimate union, and that the duality of *ousia* and *energeia* (presupposed in all language about self-mastery) is inadequate to express the goodness of the divine will which is also one with God's Being.

The Christian tradition long before Augustine had developed the doctrine of creation out of nothing. Second-century gnostic sects had sought to explain evil by saying that the creator was not possessed of the highest power, nor even of the highest goodness. Moreover, the sludge of recalcitrant matter with which the creator had to go to work was not such as to allow a satisfactory result to be achieved. The second proposition was entirely familiar in the Platonic schools. Creation was an act or process of ordering a pre-existing matter. One needed to be (like the anonymous author of the tract *On the Sublime* ascribed to Longinus) deeply sympathetic to Hellenistic Judaism and disposed to venerate the book of Genesis in order to discern sublimity in the sudden blaze of 'Let there be light'. In the second century A.D. Galen (*De usu partium* xi, 14) specifically comments on the first chapter of Genesis and criticises Moses for supposing that God can suddenly bring things into existence. Galen comments that Greeks do not think in this kind of way. He does not use the image, but the impression is clear that Galen understood Genesis 1 to imply creation out of nothing in the manner of a divine conjuring trick, producing hitherto non-existent rabbits out of (what is more) a previously non-existent hat.

The first consciously articulated doctrine of creation out of nothing appears in the Egyptian gnostic of the second century, *Basilides* (as reported in the *Elenchos* of Hippolytus). In Basilides this doctrine is a way of expressing the 'wholly other' transcendence of God, who in gnostic theology lies utterly beyond any access-route available to the human mind. Basilides' followers largely abandoned his positions, and left the ground free for orthodox Fathers to take over and to set in another framework, free of gnostic theosophy, the theme that the creator does not merely order matter, he creates it. The impetus to formulate this doctrine was reaction against the dangerous notions of the heretic Valentinus, according to whom matter was brought into existence not by sovereign act of the supreme God but as an unhappy by-product of the fall of Sophia in heaven: in short the cosmos is the consequence of a disastrous smudge, an ignorant blunder.

Against this gnostic cosmology *Justin* and *Irenaeus* put great emphasis on God's omnipotent power, whose creative initiative is seen in the birth and resurrection of Christ (*Dial.* 84 etc.; *Adv. Haer.* ii, 1:1; iii, 8:3, etc.). Neither Justin nor Irenaeus desires total rejection of the Platonic terminology. Irenaeus holds the creative act of God to be an expression of his will (iv, 38:3, etc.), and cites *Timaeus* 29e (iii, 25:5) without much misgiving. But Irenaeus does not set out to offer a philosophical reconciliation of biblical creationism with Platonic cosmology. The first systematic exposition of the Christian doctrine of creation, in which there is a genuinely sophisticated awareness of how complicated the problem really is, comes in the third century in Origen, *De Principiis*. To *Origen* it is incomprehensible that clever men (that is, the Platonists) should think it necessary to suppose that matter is uncreated and that 'creation' is a matter of God doing his best with pretty awkward and chancy stuff. Origen sees that if one accepts the doctrine of Providence (as Platonists claimed to do), there is no intellectual obstacle in the path of affirming creation out of nothing (*De Princ.* ii, 1:4; *Comm. in ev. Joh.* i, 17:203). Origen anticipates Augustine in being reluctant to accept the polarity of nature and will in the doctrine of God. He regards this polarity as a consequence of our anthropomorphic language and its inadequacy to express the being of God. Like Augustine again, he anticipates the interpretation of creation in terms of participation in the divine ideas and goodness. Both Origen and Augustine share a determination to be faithful to the biblical stress on the creative will of God, but to mitigate any suggestion that there is some element of spontaneity and suddenness about God's purposive action. Both lay great stress on the immutability of God, and on a denial that the creating of the world is an act in time. In his first homily on Genesis Origen interprets the *In principio* of Genesis 1 as meaning creation in Christ and not as a temporal event. He even uses the principle of divine immutability as a basis for urging that in dependence on the will of an external creator the cosmos itself is everlasting (*De Princ.* i, 2:10; vigorously attacked by Methodius, as quoted by Photius, *Bibliotheca* 235, p. 302a, 30ff. as prejudicing God's freedom and self-sufficiency).

The second-century pagan *Celsus* deeply objected to the Jewish-Christian doctrine of God as continually intervening to remedy the evils resulting from the incompetence of his creative handiwork; as altogether too concerned with one particular people in a small corner of the world; and as exercising even his most general providential care in too anthropocentric a manner. As a good Platonist Celsus regards interventions by the Most High as incompatible with divine immutability and perfection, and therefore has to plead that the amount of evil in the cosmos is a determined constant somehow kept in a state of equilibrium by the structure of things. Origen's reply rejects all notions of arbitrary interventions, but rests the case for biblical faith on the need for providence to hold evil in check, raising up prophets, and finally bringing the education of the human race to a climactic point in the incarnation of the divine Logos. He does not disagree with Celsus that providential care is for the whole rather than for the interests of the individual; at one point in his homilies on Jeremiah he observes that if hell is a case of deterrent punishment, it could be morally defended on just such grounds (*Hom. in Hierem.* xii, 1). He allows that what is bad for an individual may well be for the advantage of the cosmos (*C. Cels.* iv, 70). Throughout his writings Origen's theology presupposes that 'providence' is another way of speaking about divine immanence within the power and beauty of the cosmos which is not indifferent to value. Not everything that happens is the direct will of God. 'Some things occur by his will, others are by his good pleasure, but other things by his permission' (*Frg. in Luc.* 57, p. 261, Rauer). But all is ultimately absorbed within the grand design.

The Platonic tradition, whether pagan or Christian, could not easily come to terms with the fact that the same concatenation of circumstances is capable of evoking alternative responses on the part of rational creatures. Both in Origen and in Augustine there is a deep reluctance to admit that anything can happen 'by chance' if that phrase is understood to mean a wholly random occurrence. They do not ask themselves questions about the accidents of history in which the temporal coincidence of two or more altogether independent chains of events affects the outcome in ways that human foresight could not have predicted.

In such circumstances both want to discern a wise providence. In the Aristotelian tradition, on the other hand, the 'necessary' is that which is always and invariably the case, and there is a large realm which is contingent and indeterminate. The Peripatetic schoolman, *Alexander of Aphrodisias*, teaching in Athens about A.D. 200, sees Aristotle's assertion of providence in the celestial order, not in the untidy chaos of human history, as offering a middle way between Epicurean denial and the excessive assertions of the Stoics and Platonists, for whom everything that ever happens is providential. The late neo-Platonists sought to accommodate the Aristotelian tradition within a Platonic metaphysic by means of the formula, familiar to readers of Boethius' *Consolation of Philosophy*, that events apparently contingent in our eyes nevertheless fall within a divine foreknowledge. The late neo-Platonists learnt from Aristotle that 'chance' is not a cause but a way of speaking about open possibilities. Religious Platonists, both pagan and Christian, remained sure that what we do not know God does. A particular striking illustration of this confidence appears in a bizarre story told by Augustine (*De Genesi ad litteram* xii, 46-47): a horoscope cast purely as a practical joke at somebody's expense turned out in the event to be correct—and nothing could have been further from the intention of the practical jokers, who had arbitrarily invented the whole thing (or so they supposed). In the *Confessions* (ix, 3:5) Augustine similarly treats 'chance' as another way of speaking about the mysteries of providence. Human arrogance and error assumes that we can judge how providence might best act to preserve our interests. It would be more sensible to remember that the divine physician understands us and our sickness a lot better than we do (Aug. *En. in Ps.* 34:7). Heaven defend us from the armchair planners who think the world could have been much better organised than it is (Origen, *C. Cels.* ii, 68).

By prayer, Origen explains, we do not inform God or change his mind, but bring our wills into harmony with the immanent presence of God in the world where we are set. Augustine likewise (*City of God* xxii, 2) holds that praying is not putting God under pressure to alter his arrangements. Augustine rejects any notion that God ever violates his own order of nature. He does all in order, measure, and due proportion. The changing pattern of more or less regular events in our environment we call 'natural' causes. But when an unusual event occurs through which mankind experiences moral admonition and instruction, we call it a miracle. For the greater part of his life Augustine repeatedly insists that the spiritual Christian does not look for physical miracles at all but for the inner miracle of moral conversion. The counterpart of the New Testament miracles is now in the sacraments of the Gospel (*De Bapt.* iii, 16:21). The devotions of his African flock moved him to a more positive evaluation of the healings at the shrines of the saints and martyrs than he had at first held; but his principle does not vary that God's wonderful power manifest in the creation is always consistent with his particular care which is never arbitrary (*De Genesi ad litteram* ix, 17, 32-33). Christ performed miracles as nourishment for infant faith; but as faith matures, it forgoes the help of visible wonders (*De Peccatorum Meritis* ii, 32:52).

There is a consistency of pattern in the thought of Augustine about creation and providence. He does not want to make a choice between the will and the goodness of God, between transcendence and immanence, between the efficient and the final cause, but rather to affirm both to be true.

Olaf Pedersen

The God of Space and Time

1. GOD AND THE UNIVERSE

FROM THE very beginning it was an essential article of the Christian faith that the universe is created by God and by Him alone. The attempts to clarify the relations between Creator and creation led to theological investigations in two opposite directions. Firstly, following a hint by St Paul (Rom. 1:18-20) one could search the realm of creation for 'vestiges' of the Creator, or even of the Trinity, as St Augustine did not hesitate to affirm (*De Trin*. 6:10; and see *De civ*. 11:24), in order to derive some of the divine attributes from the properties of the visible world. This line of thought became a recurrent theme in later scholastic theology and was carefully developed by Peter Lombard in the *Libri Sententiarum* (I, 3:5). In St Thomas Aquinas it led to the five 'ways' (*viae*, not *demonstrationes*) along which the very existence of God is inferred from the changes, causality, contingency, order, and finality of creation (*S. Th*. 1a., Q. 2, art. 3), with the concept of analogy as the necessary intellectual tool (*ibid*., Q. 13, art. 6). Over the centuries this approach proved beneficial to science as providing a theological legitimation of the study of nature. It also provided the basis for the Natural Theologies which flourished in the eighteenth century. But, secondly, it was also possible to move in the opposite direction, using theology as an *ancilla scientiarum* by bringing Christian belief to bear upon the properties of the universe. In the following we shall concentrate (in an all too brief and sketchy way) upon this second type of approach which in some cases was used to refute theologically unacceptable cosmological ideas, and in others acted as a kind of midwife, for new and revolutionary scientific concepts.

2. PRELUDE IN THE EARLY CHURCH

Christianity emerged at a time when there was a fairly wide consensus on the physical structure of the universe. The whole cosmos was a sphere, subdivided into spherical shells and with the spherical, immobile earth at its centre. On the outside it was bounded by the *primum mobile* which rotated once a day with perfect regularity, carrying the fixed stars around, and also the planets below them with various degrees of velocity. Since the outer surface had to move with finite speed, the universe had to be of finite size. According to Aristotle space was co-extensive with matter, with the corollary

14

that outside the *primum mobile* there was nothing, not even an empty space, a point on which the Stoics disagreed. With a few exceptions the Fathers felt no qualms about the general features of this model, although they became aware of a number of problems which arose when some of its underlying assumptions were confronted with the dogma of creation.

Although St Justin (*Apol.* 1:59) and St Clement of Alexandria (*Strom.* 5:15) still believed that God made the world out of pre-existent primeval matter, the belief in a creation *ex nihilo* was firmly established at the time of Origen (*De prin.* 1:3). Because of the Aristotelian connection between matter and space it followed that space itself was created together with the material world. Before the creation there was no space. It might then be asked 'Where was God before heaven and earth existed?'. Augustine's answer that 'God dwelt in Himself and by Himself' (*In Ps.* 122:4) was another way of saying that God is without spatial or dimensional attributes. Nevertheless, it proved difficult to speak of the omnipresence of God without using spatial terms. Arnobius called God 'the place and space of things created' (*Adv. haer.* 1:31), and in Augustine we find the often quoted metaphor of the world immersed in and penetrated by God's infinity as a sponge soaked in an immense sea (*Conf.* 7:5). In a less metaphorical language he stated that God is 'wholly in the heavens, wholly on the earth, wholly in both, and not confined to any place' (*Ep. 187 ad Dard.*), just as he affirmed that if there were a void beyond the heavens (which he did not believe) it, too, would be filled with the essence of God (*De civ.* 11:5).

Defining time as 'that by which motion can be measured' Aristotle had connected time as closely with motion as he had tied space to matter. In itself this proved acceptable to Christians who, nevertheless, had to object to two corollaries to this definition. In the first place Aristotle's prime indicator of time was the rotating firmament as the great natural clock of the universe. Being unable to find physical causes which might slow this rotation down, or speed it up, he assumed it to be everlasting with the consequence that the universe was eternal. Now Holy Scripture revealed in its very first sentence that the world had a beginning. Here Christian belief was clearly at variance with the prevailing cosmology, although not with the Aristotelian definition of time as such. As Augustine explained, since motion and time belong to matter 'Time could not have been before some moveable thing was created' (*De civ.* 11:6). In other words, time as well as space was created together with matter, and there was no time 'before' creation. On the other hand, it proved impossible to find a rational proof that time had a beginning. This belief rested completely on the authority of Scripture, which also made it possible to calculate the actual age of the world. In the end Aquinas had to admit that no proof was possible and that creation in time was a truth held by faith alone—*sola fide tenetur* (*S. Th.* 1a., Q. 46, art. 2).

The belief in created time naturally led to the question 'What was God doing before the world began?' as reported by Origen (*De princ.* 3:5), whose own answer reveals the second element of the doctrine of time which troubled the theologians. Time was not only eternal—it was also cyclical. This idea was of astrological origin and based on the belief that all events are caused by the influence of the heavenly bodies. Consequently they must happen in exactly the same way when, after the elapse of a 'Great year', the planets return to the same mutual positions (*C. Celsum* 4:67) as believed by both the neo-Platonists following the *Timaeus* and the Stoics who assumed an eternal succession of identical worlds separated by universal destructions. Origen agreed and answered the question by saying that before this world began God was busy with a previously existing world, just as there will be another world after the dissolution of the present universe (*De princ.* 3:5). Even if he tried to find support for this view in Scripture (Isa. 66:22; Eccl. 1:9f) it was resolutely rejected as incompatible with the uniqueness of the Christ-event and the belief in the kingdom of God. 'God forbid that we should believe this', said Augustine, 'for Christ died once for our sins, and rising again, dies no more', just as we after our resurrection 'shall always be with the Lord' (*De civ.* 12:13; cf. Rom. 6:9 and 1 Thess. 4:17). This appeal to the belief, not only in creation, but in redemption and salvation, proved strong enough to eradicate the notion of a cyclical time from theology. The answer to the question was worked out in terms of the non-temporal 'eternity' of God which differs from created time by not being a succession of passing instants, as Augustine explained (*Conf.* 11:11-13). Later Boethius gave the classical definition of eternity as the 'complete possession of unending life all at once' (*De cons. phil.* 5:6). The result was that, until the coming of evolution in the nineteenth century, it was the problem of God and space which dominated the discussion.

3. THE PROBLEM REVIVED

The Patristic discussion of the relations between the belief in God and the notions of space and time was in many ways only a prelude to a more far-reaching debate which flared up after the middle of the thirteenth century. By this time the general knowledge of ancient philosophy had vastly increased. Aristotelianism had acquired a safe position in the curriculum of the universities where the moderate and essentially independent attitudes of Albertus Magnus and Aquinas were challenged by a radical school of extreme Aristotelians leaning on the authority of Averroës. But the new translation of Aristotle's work completed in 1269 by William of Moerbeke, at the request of Aquinas, separated Aristotle from his Arabic context and made an independent study possible. At the same time alternative cosmological ideas became known. Moerbeke also translated Simplicius' commentary on the *De caelo* which revealed that the Stoics had ideas of their own. They did in fact assume the existence of a void outside the heavens for 'if someone standing motionless at the extremity of the world extends his hand upwards' he will actually reach into outer space—an assumption denied by orthodox Aristotelians for the reason that there was no space out there to receive the hand, as Aquinas agreed in his subsequent lecture on this question (*In libr. De caelo*, I, 21:209). Also the old Latin translation of the Hermetic dialogue *Asclepius* gained new interest by asserting that 'not even the region outside the cosmos is void, seeing that it is filled with things apprehensible by thought alone, that is, with things of like nature with its own divine being' (ed. W. Scott, *Hermetica* I, 1924, 319 ff), a view which enabled 'Hermes' to uphold the denial of a void in an infinite world since everything is filled with either matter or spirit.

The event that triggered the new debate was the condemnations in 1270 and 1277 by Bishop Etienne Tempier of Paris of two lists of heretical propositions allegedly taught in the University. Among them were the doctrines of the eternity of time and matter as held by the Averroists. New was the formal condemnation of the Aristotelian doctrine of necessity as represented by propositions like 'The human will wills and chooses of necessity' (1270 no 3) with its astrological counterpart 'Everything that happens in the lower world is subjected to the necessary influence of the heavenly bodies' (1270 no 4). It was followed by a condemnation of the cyclical view of history as expressed by the proposition 'When all the heavenly bodies return to the same point, which takes 36,000 years, one will observe the same effects as now' (1277 no 6). Space is referred to in the statement 'God cannot move the heavens along a straight line for the reason that then it would leave a vacuum behind' (1277 no 49)—a sentence which has puzzled scholars since no Aristotelian would ever consider the possibility of a translation of the universe since they had no extra-mundane space in which to move it. However, it makes sense if taken as a reference to the 'new' Stoic doctrine of a void beyond the world. Finally a number of purely theological heresies, such as 'God does of necessity everything which stems directly from him' (1277 no 53), or 'God cannot be the cause of something whose existence has a beginning' (1277 no 48) and 'God cannot produce the effects of a secondary cause without the help of this very cause' (1277 no 63). In spite of many obscure details the general purpose of this intervention clearly was to protect the belief in the freedom and omnipotence of God in relation to a non-necessary, contingent world.

4. TOWARDS AN INFINITE UNIVERSE

The many references to this condemnation during the next hundred years reveal that philosophers now had a new and powerful tool for attacking traditional views of the universe. Such views might be right; but they were not necessarily right since God could have arranged things in a different way. Phrases beginning with 'God cannot . . .' had to be regarded with suspicion. The most consistent result of this change of attitude was, perhaps, the theology of William of Ockham. But more important in cosmology were the new ideas put forward in 1344 by Thomas Bradwardine in the first book of his huge work *De causa Dei* (1:5). Although primarily directed against Pelagian trends in contemporary theology and ethics it also contained such cosmological propositions as (1) 'God is everywhere in the world and in all its parts'; (2) 'And also beyond the real world in a place, or in an imaginary, infinite void'; (3) 'And so truly He can be called immense and unlimited'; (4) 'And so a reply seems to emerge to the old question of the gentiles and the heretics—"Where is your God?" and "Where was God before [the creation of] the world?" '; and (5) 'And so it seems obvious that a void can exist without a body, but in no way without God'.

This first affirmation of an infinite space in Christian thought may be inspired by the *Asclepius* which was known to Bradwardine. But his real basis was the condemnation of 1277 which had opened his eyes to the fact that by denying the existence of such a space (in which the world might be moved) the Aristotelians 'seriously diminish and mutilate the divine, indeed omnipotent power' of God (*ibid.*).

From now on the traditional academic *quaestio 'Utrum aliquid sit extra coelum?'*—Whether there is anything outside the heavens—appeared in a new light. It could be answered in more ways than one since Bradwardine had not explained why he called the extra-mundane space 'imaginary'. Since he stressed that God is 'infinitely extended without extension and dimension' it may be that the term 'imaginary' was used as a safeguard against the ascription of dimensions or other spatial attributes to God. Another interpretation appeared already in 1354 in the commentary on the Sentences (1:37) by John of Ripa who provided the void beyond the heavens with the three ordinary dimensions at the same time as he gave it a different and lesser degree of 'immensity' (i.e., immeasurability) than that possessed by God. A third proposal was due to Nicole Oresme who in his French commentary on the *De caelo* (1377) maintained that the imaginary 'infinite and indivisible space is the immensity of God and is God Himself' (1:24). It is different from ordinary space just as God's eternity is different from time. 'Imaginary' seems here to mean that we can infer the existence of infinite space by reason, without experiencing it, just as we can infer God's eternity, but only experience time.

Thomas Bradwardine also referred to another line of thought by quoting the pseudo-Hermetic *Book of the 24 Philosophers* to the effect that God is 'an infinite sphere whose centre is everywhere and circumference nowhere'—a definition which in 1440 was adopted by Nicholas of Cusa as one of the guiding principles for the new cosmology he sketched in the *De docta ignorantia* (2:15). Here the universe is unlimited (if not infinite?) and everywhere filled with stars among which is the moving earth, conceived as a 'noble star' among others. From each of these bodies the universe would appear to an observer in very much the same way—an amazing intuition of what has later been called the general cosmological principle.

5. TOWARDS A SPATIAL GOD

In the two following centuries the new cosmological ideas concerning space were transmitted through two different channels. One of these originated in Cusanus and led to the eclectic, neo-Platonic and Hermetic speculations of Marsilio Ficino, Giordano Bruno, Robert Fludd, and others. Here we shall follow the other channel through which the ideas of Bradwardine seem to have become more or less common property among post-Tridentine theologians, particularly in Spain. The great set of commentaries on Aristotle published in the late sixteenth century by the Jesuits at Coïmbra were clearly marked by his conception of space, and that was also the case of the *De caelo*-commentary of the Alcalà theologian Franciscus Murcia de la Llana (Madrid 1615) which taught that the man at the extremity of the spheres would be able to extend his hand into an imaginary, but not real space outside the heavens—a space in which God is just as present as in the finite, material world (3:10).

It goes almost without saying that it became increasingly difficult to uphold the obscure distinction between 'imaginary' and 'real' space. This was the opinion of Franciscus Patrizi who, in his *Nova de Universis Philosophia* (1591), took the decisive step of equating the two kinds of space, regarding space of any kind as the first of God's creatures and therefore real, at the same time as it was actually infinite. This was a conclusion which would have been unacceptable to the scholastics, since it seemed to locate God's 'immensity' in something created by Himself. But the most powerful reason for abandoning the distinction appeared in 1672 with the publication of the *Experimenta Nova* by Otto von Guericke, who has become famous as the founder of vacuum physics by his invention of the airpump, by which he proved that it was possible to create an extended vacuum inside the material world. Thus one of the essential tenets of Aristotelian physics had to yield to the force of the experimental method. But it is interesting to notice that Guericke also rejected the Aristotelian denial of outer space with explicit reference to the Coïmbra Jesuits, just as he refused to accept the Cartesian idea of the identity of matter and extension. His conclusion was that 'the "nothing" beyond the world, and [ordinary] space are one and the same, and so-called imaginary space is true space' (2:6). Nevertheless, Guericke refrained from ascribing dimensions to outer space because he, like Oresme, still regarded it as divine (2:8).

At this stage there had to be a parting of the ways. One possibility was to preserve the connection between space and a non-dimensional God by denying the dimensional character of infinite space. Another was to ascribe dimensions to this new kind of space (under the influence of vacuum physics) at the cost of making God a dimensional being. A third possibility was to sever the connection between God and space altogether.

6. AROUND NEWTONIAN PHYSICS

It seems that the second possibility was adopted by the Cambridge Platonist Henry More who, in his *Enchiridion Metaphysicum* (1671), described the infinite void as a three-dimensional container for all movable things in the world. As such it is not only absolute (i.e., not depending on the contained bodies for its existence), but also possessed of a number of other 'titles' listed by More—such as immensity, eternity, omnipresence, and similar attributes which to a large extent corresponded to the traditional attributes of God. The final results were, firstly, that space appeared as an attribute of God and, secondly, that God became a dimensional being and therefore no longer able to be fully present in any finite part of the universe—a conclusion which in 1702 was worked out in great detail by the mathematician Joseph Raphson.

A look at Newton's *Principia* from 1687 shows the extent to which the father of classical physics was indebted to Henry More. In fact, Newton's conception of 'absolute space' is almost indistinguishable from More's three-dimensional container space, and Newton's 'absolute time' is from this point of view just the logical counterpart to space. We shall not here go into the reasons why Newton found it necessary to base his mechanical system on these two concepts, but only point to the fact that Newton was aware of their theological implications although in the *Principia* he omitted any reference to theological points of view. However, in the Latin version of his *Opticks* (1706) he inserted a number of new 'Queries' which seemed to imply that he regarded physical space as, as it were, the 'boundless, uniform Sensorium' of God. If this meant that God was dependent upon space for his knowledge of the created world it would be a radical departure from the traditional concept of God as transcendent and essentially independent of creation.

The reaction was not long delayed. In his *Principles of Human Knowledge* (1710) Berkeley drew the conclusion from Newton's ideas that 'either real space is God, or else there is something besides God which is eternal, uncreated, indivisible, immutable' (§ 117). But the main attack came in 1715 in a letter to the Princess of Wales from Leibniz who, among other things, complained about the growing ungodliness in England where philosophers ascribed material properties to God. This gave rise to the famous correspondence between the German philosopher and Newton's faithful disciple Samuel Clarke which cleared up the difference between Newton's notion of space as an absolute entity independent of matter, and Leibniz's notion of space as a relative concept closely related to matter. During the debate, however, it became increasingly clear that the issue was no longer how the properties of space and time might be derived from the attributes or concept of God. The very concept of God itself was now in the process of being coloured or changed in one way or another by the different philosophies of nature advocated by Leibniz and the Newtonians. In this respect the debate transcends the subject of this paper. It belongs less to the tradition moving from God to the world than to the new Natural Theology of the eighteenth century in which the intellectual movement went in the opposite direction. Here it was once again a question of obtaining knowledge of God from 'the marvels of creation'. The names of Paley and Whewell and many others testify to the vitality of this tradition which was not checked until the theory of evolution re-opened the debate on God and time which had been almost overshadowed by the problems of space.

That Newton himself was instrumental to this shift of emphasis appeared in his previous correspondence with Richard Bentley whose printed version of his Boylean Lectures on *The Confutation of Atheism* (1693) owed some of its principal ideas to Newton. In fact, Newton told Bentley that because of the universal attraction between bodies the universe would suffer what is now called gravitational collapse unless God directly and all the time counteracted the effect of gravitation by what in 1694 he described to David Gregory as a 'permanent miracle'. This divine intervention would ensure the stability both of the universe in general and of the solar system in particular—a line of thought which was not broken until one century later when Laplace offered a proof that 'this hypothesis' was unnecessary for the stability of the solar system. The stability of the universe as a whole is, scientifically speaking, still an open question.

CONCLUSION

Strictly speaking there can be no conclusion to a purely historical account. On the other hand, one has to admit that the long history which has been described here in its briefest outline seems to indicate two points of general interest. One is the fact that Christian belief and doctrine at one stage (in the wake of the events of 1277) served as a handmaiden of science, helping to liberate man from the bonds of the old cosmos and to let him loose in that infinite abyss which filled Pascal with horror. Another is that this attempt to 'make use of' God had to be abandoned again at the moment when a new conception of the universe threatened to destroy that concept of an omnipresent, but non-dimensional God which had helped it to emerge. In the end neither theology nor scientific cosmology remained unaffected by this strange encounter. Where theology is concerned it seems impossible that we shall ever again speak of the omnipresence of God in a language derived from purely scientific considerations of the universe.

Bibliography

It is impossible here to give references to the editions of the many works quoted by chapter and verse in this paper. The following titles are essential to the study of the subject.

A. Koyré 'Le vide et l'espace infini au XIVe siècle' *Archives d'histoire doctrinale et littéraire du moyen âge* 24 (1949) 45-91 (On Bradwardine; texts in French translation).

A. Koyré *From the Closed World to the Infinite Universe* (Baltimore 1957) (Deals with the period from Cusanus to Newton).

E. Grant *Much ado about nothing: Theories of space and vacuum from the Middle Ages to the Scientific Revolution* (Cambridge, Mass., 1981) (Covers the same ground as Koyré, but in much greater detail and with more theological acumen. The principal work on the subject).

E. Grant (ed.) *A Source Book in Medieval Science* (Cambridge, Mass. 1974) (Contains the relevant texts by Bradwardine, Oresme, the Coïmbra Jesuits, Guericke, and others in transl.).

R. Hisette *Enquête sur les 219 articles condamnés à Paris* (Louvain-Paris 1977) (Philosophes Médiévaux, XXII).

P. Duhem *Le Système du Monde*, VI (Paris 1954) (the aftermath of the condemnation).

S. Jaki *Science and Creation: From eternal cycles to an oscillating universe* (Edinburgh and London 1974).

H. G. Alexander (ed.) *The Leibniz–Clarke Correspondence* (Manchester 1956) (Annotated edition of the texts).

J. E. McGuire 'Existence, Actuality and Necessity: Newton on Space and Time' *Annals of Science* 35 (1978) 463-508.

Günter Altner

The Evolutionary Shift

BECAUSE OF the changes it brought about in the ways in which we look at the world, Darwin's theory of evolution has been compared to the upheaval initiated by Copernicus. Is such a comparison justified? It is hard to say. What sort of criteria would be applicable? His contemporaries' sense of personal shock? The effects on the various disciplines and world views caused by the particular theory? Such comparisons are on shaky ground. Yet there is much in favour of the adjective 'Copernican' as applied to the change wrought by Darwin's theory. It did in fact change our world view, both inside and outside biology. And his contemporaries did experience it in this way. If we look at eighteenth century biology and natural philosophy, in particular at the end of the century, we are surprised at the degree of upheaval after 1859. For in the eighteenth century there were many traditions which could have linked up with Darwin's theory. On the other hand we have to deal with the fact that a deep trauma did happen to the world view, and that it called forth strong opposition, especially in Church circles. Only by accepting both aspects—continuity and the new departure—can one properly evaluate what is distinctive and specific in the revolution started by Charles Darwin.

There was a great deal of controversy even in biology at the time. Some gave Darwin's theory an emphatic welcome, hailing it as a prophetic and integrating force in biology. No doubt August Weismann spoke for many in praising the *Origin of Species* as a 'book of revelation' which had given him a new sense of direction in working with his wealth of detailed research: 'Many felt the same. A whole generation had arisen that was tired of speculation, and from the beginning of the century they rushed into the arms of specialist disciplines, aiming only at establishing new facts. Now Darwin supplied the link concept: evolution.'[1] Those who thought and felt like this were biologists and medical men, in particular those scientists who were repelled by the speculative natural philosophy of the eighteenth and nineteenth centuries and enthusiastically went in for the classification studies in zoology, botany and comparative anatomy started by Linnaeus (1707-1778). They pursued an empirical classification of the objects of the animal and plant worlds, extending it to the total function of animal and plant bodies. This threatened to get out of control, but it was here that Darwin's theory of evolution had a systematising effect. On the other hand, there was also resistance to Darwin's theory within the field of nature research, particularly with regard to the conclusions which were being drawn concerning the origin of man. In 1864 210 English natural scientists, including David Brewster, Balfour and others, voted against Darwin and Huxley. They insisted that with regard to the origin of man there must be no contradiction between the divine revelation in the book of nature and that in Holy Scripture. Therefore it was a question of the implications of Darwin's attempt at an explanation in relationship to the biblical idea of creation.

However, these opposing positions within the research community for and against Darwin must not obscure the larger context. These reactions resulted in each case from the particular mentality of the individual concerned—an empiricist inimical to speculation, or a conservative believer—and represent only isolated views within a much wider spectrum of opinions and schools of thought. Looking backwards from the years of the dispute around 1859 one can see that all the biologists of renown at the start of the nineteenth century were also natural philosophers; the anatomist and embryologist L. Oken, the geologist H. Steffens, the physiologists J. Müller, K. Burdach, J. Purkinje, the morphologists J. B. Spix, K. G. Carus, the anthropologist J. F. Blumenbach, the botanists Nees von Esenbeck, A. von Braun, the embryologist C. E. von Baer and others. But in France and England, too, there were adherents of this especially German 'natural philosophy' approach, for instance Geoffroy-St. Hilaire, Blainville, R. Owen and the Dane, Oersted.

The idea of evolution had been familiar to this range of 'nature philosophy' biologists long before it received its particular shape from Darwin, albeit often in the garb of related concepts like transition, progress, metamorphosis, perfection. This shows the connection with the philosophy and in particular the philosophy of history, of the late eighteenth century. Here people had been trying for a long time to understand man's historicality and the historicality of nature as partial phases of a whole. In his *Ideas on the Philosophy of History of Mankind* of 1784, J. G. Herder meditates on the 'realm of human organisation' within the total field of creation: 'If we may be permitted to make some assumptions about the hidden forge of creation, the human race can be seen as the great confluence of lower organic forces which are intended to result in the formation of humanity in him. And what follows from this? Man bears the image of the divinity and enjoys the most refined organisation the earth could give him; is he now to go backwards and become again a tree trunk, a plant, an elephant? Or has the wheel of creation stopped for him. . .? Let us look back and see that all that has been seems to find its maturity in the figure of man, and that with regard to man himself what we see is only the first bud, the initial stage of what he is to be, of that for which he was designed. . . .'[2] In his writings on anthropology, the philosophy of history, politics and education, Kant, at the end of the eighteenth century, comes to conclusions similar to Herder. In the history of the human genus Kant discerns nature's design of leading man out of the 'gnarled wood' of the 'lower stage of animality' into the realm of freedom in the form of 'civil society'.[3] In the case of Herder in particular we can see how big a part the Christian conception of time and history plays in the philosophy of history of the eighteenth and nineteenth centuries and in the idea (which it presupposes) of a goal-oriented evolution. On the basis of a definite teleology in God's saving dealings with the world, nature and history are joined together in a unity which still awaits its perfection. The dynamism of this concept of history, which was able to see body and spirit, instinct and freedom, as concomitant and alternating configurations through time, was far in advance of many of the barriers erected against Darwin's theory, far in advance of the theory that the species were fixed on the first day of creation, far in advance of the (ultimately Hellenic) suspicion of the body found in many theological thinkers, and far in advance of all those theories of evolution which understood it by way of pre-formation merely as the unfolding and growth of pre-established micro-structures. According to this view, which was also supported by Leibniz, one would have to suppose that 200,000 million human beings had been packed inside Eve, or even in Adam's loins. This view was also shared by the great physiologist Albrecht von Haller in the middle of the eighteenth century. But Herder, who, with the embryologist C. F. Wolff, started from the realisation that the organism in the ovum (or in the sperm) was not pre-formed, spoke of formation by 'organic forces': 'No eye has even seen pre-formed embryos, ready-made since the creation; but what we do see from the first moment of a creature's life are organic forces at work . . . a formative process (genesis), the effect of inner forces, prepared by nature in masses. . . . In the deepest abysses of the process of becoming, where we see the very germs of life, we become aware of an unexplored region, an active element, to which we give the inadequate name of light, aether, the warmth of life. It may be that this is the sensorium of the All-creating One, by which he enlivens everything and gives it warmth.'[4]

These ideas of Herder link us once again with the natural philosophy of leading biologists at the beginning of the nineteenth century. Among them was a considerable number of researchers who were able to express in a suggestive and theoretical way the historicality of the natural world and its changeability through time. Mention must be made in particular of the works of G. L. L. de Buffon (1707-1788), J. B. de Lamarck (1744-1828) and E. Geoffroy-St. Hilaire (1772-1844). The grandfather of Charles Darwin, Erasmus Darwin (1731-1802), also deserves mention. In his work *Zoonomia* he had given a poetic, pantheistic turn to the idea of evolution. In general we can say of this group of Darwin's predecessors that they set out from a general conception of the animal (and plant) world and express this assumption of a gradual development of natural life-forms in a more or less direct manner.

With Lamarck this suggestive approach comes most clearly to fruition. He inquires about the 'forces' which produce the ladder of natural forms from the simplest to the most complex beings. In explanation Lamarck points to the inner needs of the animal body which are transmitted, as a result of reinforcement, to the next generation as bodily characteristics. At the same time Lamarck sees warmth, light, electricity and moisture as genuine factors of external nature, contributing to the emergence and development of the organisms.[5] In tracing the ladder of organisms back to the effects of warmth, light and electricity, Lamarck connects up with Herder, who referred to light, aether and the warmth of life as the 'sensorium of the All-creating One'. On the other hand this view of the situation shows that we are at the very centre of those physiological speculations current among the 'natural philosophy' biologists at the turn of the eighteenth/nineteenth century. Naturally, too, we are in the area of Goethe's *Metamorphosis of Plants*, the influence of which must not be underestimated in this connection. For Goethe, the structural principles of plant and animal and their variations in the development of embryo and phylum were tangible concretions, even revelations of what was really at work within the changing natural forms: the spiritual. Linking up with his successful morphological studies in botany, his discovery of the human intermaxillary bone and his theory of the skull's development from the spine, Goethe also goes back to speculative, metaphorical concepts such as light, air, attraction and repulsion, in order to explain the creative power which can change matter. He also puts forward the idea of metamorphosis in a way which relativises all formal characteristics and all definitions; they are inadmissible, rigid formulas, incapable of expressing the changing forms which are incessantly transcending themselves: 'If we are to think in terms of a morphology, we must not speak of "form"; or rather, if we do use the word it should only represent the idea, the concept or some empirical detail momentarily held fast. What is formed is immediately transformed, and if we hope in some degree to behold nature in its living reality we must keep ourselves in a similarly fluid and malleable state.'[6]

This was the shape of much thought on the dynamism of natural development at the turn of the nineteenth century. The development of forms was to be subject to speculative understanding, not to experiment. The first to summon the imagination as a suitable faculty for understanding the metamorphosis of forms was the famous physiologist Johannes Müller (1801-1858), the teacher of Ernst Haeckel. It was the speculative mind, an observation of forms which aimed at a philosophical understanding, which was to reveal the deeper mystery of the development of life-forms through time, not the 'unreliable experiment'.[7] Thus, decades before the publication of Darwin's *On the Origin of Species* (1859), the idea of evolution was taken as a matter of course among the élite of German and European biologists. But it was interpreted as the mode of appearance of 'the spirit', not yet complete, in the many-layered and increasingly complex realm of natural forms (right up to and beyond recent man). Not only was the future of this development open; its occurrence meant that the hour had struck for the ultimate unveiling of the forces at work in each life-form. And then Charles Darwin came with his theory of evolution.

Again and again in his books Darwin stresses the existence of a general evolutionary connection between the species and tries to demonstrate this by adducing a great deal of evidence. In this he is close to the champions of evolutionary thought in the late eighteenth century, although he does not take up their observations directly. 'There is grandeur in this view of life,' Darwin observes at the end of the first edition of *On the Origin of Species*, 'with its several powers, having been originally breathed into a few forms or into one; and that, whilst this planet has gone cycling on according to the fixed law of gravity, from so simple a beginning endless forms most beautiful or most wonderful have been and are being evolved.'[8] In his later publications Darwin did not discount an openness of the 'organic ladder of ascent' with its implied possibility of a 'higher destiny', particularly in the case of man.[9] These aspects, too, manifest a certain closeness to the 'natural philosophy' developmentalists of the turn of the century. But here the paths divide. The primary purpose of Darwin's discussion of man's origin was to establish its consonance with the characteristics and behaviour of the 'lower forms' and 'lower animals'. It was left to Darwin's pupils and friends, up to and including Ernst Haeckel, to describe man as a product of evolution from the lowest animal realm. Here the subtle interpretation of the process of natural development as the progress of the spirit within nature was turned upside down: man's spirituality was relativised by comparative biology and ultimately razed to the ground by materialism. People no longer saw in the moving stone, the planet, the animal, or in human consciousness anything but mere further instances of a universal process and its mechanical laws. This also destroyed the historicity of nature, the temporal form of the evolutionary process striving forward to higher realisation, which Darwin had only just rediscovered and brought to light. Characteristically it was Ernst Haeckel who passionately attacked the Second Law of Thermodynamics (the law of entropy), the only law of classical physics to presuppose the irreversibility of time. For him there was only one eternal and immutable law, which he identified with the laws of nineteenth century mechanical physics. The mechanical theory of evolution was born!

It was a paradox: Darwin's concept of evolution did not lead in the end to an opening-up of process ideas in nineteenth-century biology, although it could have done; instead it led to an a-historical, mechanistic rigidity. The mechanistic theory of evolution then proceeded to obliterate the special position of man in nature. Now man was no longer a special witness to the progress of the spirit in nature (the idea of metamorphosis no longer met with any understanding at all), but at best a prominent piece of evidence for the universal effect of the eternal law of matter. In this context, certainly, the upheaval brought about by Darwin has indeed a Copernican character. Once and for all it broke the anthropocentric Western view of the world which had remained intact since the middle ages. The 'natural philosophy' developmentalists of the turn of the nineteenth century had been able to maintain man's special position by interpreting it as nature's first step upwards out of instinctual animality into the realms of freedom. All this was now gone.

The theologian Rudolph Otto saw the weakness of 'Darwinism' primarily in the way it suppressed the idea of a development oriented towards man's perfection: 'Only by being anti-teleological does it finally become expressly anti-theological.'[10] This was the view of many theologians and philosophers, particularly those who continued to be attached to the developmental ideas of nineteenth-century natural philosophy. Otto's polemic takes issue with a specific point in Darwin's theory. In fact Darwin's real achievement was not the theory of evolution. As we have seen, it had its antecedents. Darwin only dared to make his theory of evolution public when he thought he was able to provide an immanent scientific explanation for evolution in his theory of natural selection. This was where the demythologising of the purposeful creator first took place, and where biology overtook all the 'spirit' concepts of development. In this connection Darwin spoke of the 'united effect and product of many natural laws',[11] and in particular he had in mind five factors: (1) Inherited variability; (2) The formation of a surplus of descendants; (3) The struggle for survival; (4) The elimination of surplus descendants and the survival of the fittest as a result of better adaptation through genetic mutation; and (5) Change in the characteristics of the species in the direction of better adaptation.

In the public debate the complex of factors so carefully analysed by Darwin has continually been reduced to the two major factors 'heredity and selection' (the struggle for survival). Be that as it may, for Darwin and his contemporaries the breakthrough consisted in his having seen that the 'struggle of nature' and the factors behind it could result in 'ever higher and more perfect beings', including man and his spiritual achievements. Instead of the mysterious 'generative ferment of the spirit' (Johannes Müller) which was invoked by the eighteenth-century physiologists and natural philosophers, Darwin proposed material factors, in particular heredity and selection, acting aimlessly according to a pattern of chance and necessity. On the one hand, this ushered in the materialist appropriation of the theory of evolution, stylising the patterns of activity described by Darwin as eternal and immutable laws of matter. On the other hand, this is where modern evolutionary biology began, which is now attempting to submit the evolutionary process on earth to deliberate control, on the basis of molecular biology and genetic technology. No doubt there is great danger here.

Meanwhile, scientific analysis of the factors discovered by Darwin has shown that Darwin's idea of competition (the struggle for survival) owes something—consciously or not—to the concept of competition of 'Manchester' liberalism (Malthus and Spencer).[12] In spite of this it remains a fact that earthly nature can be manipulated according to the factors drawn from Darwin's theory, corroborating his attempt at an explanation. Because of these successes, people have overlooked the fact (and still do) that Darwin's theory of evolution genuinely bears witness to man's special position and to the cognitive achievements of which he alone is capable. Only a human being is able to react to himself and to his own natural history in the way every evolutionary biologist since Darwin does. However, this reflection on the preconditions of human knowledge does not allow us to take Darwin's theory of evolution as a description of nature as such, a timeless nature. Rather, the theory is a many-sided model of nature, useful and subject to abuse, an interpretation of man who is situated within the process of evolution, a product of eighteenth- and nineteenth-century thought. But far from clearing up the mystery of the interplay of nature and spirit in the world's coming-to-be, ever present in eighteenth-century natural philosophy, it only re-invokes it. Now, part and parcel of time's irreversibility, we are rediscovering that human knowing is no more and no less than a participation in the process of the universe. Knowing this, we are

C

travelling towards a tomorrow which is no longer contained within today. The inevitability of mechanistic thought is broken. Evolution is not the eternal return of the same.

Translated by Graham Harrison

Notes

1. Quoted in *Der Darwinismus–die Geschichte einer Theorie* ed. G. Altner (Darmstadt 1981) p. 474.
2. J. G. Herder *Ideen zur Philosophie der Geschichte der Menschheit* (Darmstadt 1966) p. 138.
3. I. Kant *Schriften zur Anthropologie, Geschichtsphilosophie, Politik und Pädagogik–Erster Teil* (Darmstadt 1970) p. 40ff.
4. J. G. Herder, in the work cited in note 2, p. 134f.
5. J. Lamarck *Zoologische Philosophie* (Stuttgart 1909) p. 68ff. See G. Altner, the work cited in note 1, p. 83ff.
6. J. W. Goethe *Werke* vol 13 (Hamburg 1962) p. 55f.
7. Cf. E. Rádl *Geschichte der biologischen Theorien in der Neuzeit* vol 2 (Hildesheim/New York 1970) pp. 31ff., 75ff.
8. C. Darwin *On the Origin of Species* (A Facsimile of the First Edition, Cambridge, Mass. 1964) p. 490.
9. C. Darwin *The Descent of Man and Selection in Relation to Sex* (2 vols. London 1874).
10. R. Otto *Naturalistische und religiöse Weltansicht* (Tübingen 1929) p. 107.
11. See C. Darwin *On the Origin of Species, passim.*
12. See G. Altner, the work cited in note 1, pp. 53ff., 95ff.

Tshibangu Tshishiku

Eschatology and Cosmology

CHRISTIAN ESCHATOLOGY is a doctrine about the last things, the circumstances of the fulfilment of the individual human being and of humankind, of the definitive establishment of the kingdom of God. It is a doctrine with a history during which points of view have evolved and different aspects have been emphasised in turn. Eschatological doctrines derive from revelations in certain religions, and in all cases they reflect, at least to some extent, the cosmological imagery of a given time and civilisation.

Cosmology is nothing other than a general framework for the intellectual and spiritual organisation and integration of the world in its totality, within which human beings locate themselves in relation to other beings and set up a system for interpreting their lives, with particular reference to their origin, present development and final destiny.

Even a Christian eschatology is marked or formed by the cosmological ideas dominant in each period. Since the beginning of Christianity three main cosmological schemata can be distinguished. The first was associated with Jewish apocalyptic and later influenced by Greek ideas. The second was the product of the Copernican revolution (1543), perfected by the celestial mechanics of Newton (1642-1727), and the last is that of the nineteenth and twentieth centuries, marked equally by the discoveries in biology and in the physical sciences.[1]

In addition to the effect of scientific discoveries, we must recognise the influence of developments in the human sciences, the sense of history, philosophy, literary and artistic movements, the social sciences.

In this article we shall summarise the basic features of the Christian eschatology associated with a cosmological view which was on the whole static before outlining the positions adopted and the questions discussed since the end of the nineteenth century. Finally we shall note new eschatological perspectives and indicate directions of study which may enrich eschatology through the findings of comparative religion.

1. GENERAL FEATURES OF ESCHATOLOGY IN A FIXED CONCEPTION OF THE WORLD

(a) According to the traditional teaching, the death of a human being, the effect and the 'wages' of sin, constitutes the supreme event which ends the period of a person's life, the period during which the person, by his or her actions, commitment and attitude to

27

God, who has decided to save them in and through Christ, may merit or fail to merit salvation for eternity.[2] At the moment of death the moral and spiritual balance of a human life is drawn and so the individual's eternal fate is fixed. Despite this theologians speculate about whether a person is not given, at the very moment of death, at the last moment of life, the opportunity to review all the moral and spiritual choices of his or her life and so enabled at the moment of death to make a final choice for or against God's salvation.[3]

Once death has occurred, the person appears before God's judgment, which takes place immediately.[4] The recompense due for the total of a person's deeds, reward or punishment, is then declared.

In theory the soul of the person found pure and perfect is admitted immediately to the presence of God in heaven. Heaven is described both as a 'place' and as a 'state'. Theology has defined it as the place of perfect happiness, of the bliss which consists in nearness to, and the direct, intuitive and beatific vision of, God.

In fact, in most cases souls, being stained by at least minor impurities, by what are called venial sins, are not allowed to enter heaven immediately. They pass through a preliminary stage of purification known as 'purgatory'.[5]

On the other hand, when the dead person's soul is found to be stained with serious moral faults, mortal sin, a 'radical rejection of God', it is punished with hell. This is the place of the total punishment which consists in the eternal and absolute privation of God.[6]

On this level of the ultimate fate of individual human beings, fundamental questions have always been raised:

(i) about the judgment: If the judgment which takes place immediately after the death is in itself definitive for salvation or damnation, what is the role of purgatory and the value, proclaimed by the tradition of the Church, of intercessory prayer by the living for the dead?

(ii) about hell: Can this punishment, absolute in its nature and infinite in time, be conceived as applying to a creature as finite, imperfect and limited in his actual nature as man? The Church, however, has always maintained it as an indubitable element of revelation.[7]

(iii) another question concerns the state of the soul waiting to rejoin the body at the resurrection at the end of world history. In what exactly does the 'immortality' of the soul consist, and what is its basis, 'material' or 'spiritual'?

This last question leads us to a consideration of the ultimate fate of humankind as a whole.

(b) The return of Christ, his 'parousia', is looked for as the event which will bring to an end the history of the world and inaugurate God's new day (Rev. 21:23). As the profession of Christian faith proclaims, Christ will then come to 'judge the living and the dead'.

Christ on his return will deliver the last judgment, which will consist of a general assessment of the history of the world and of humankind, the total revelation to creatures of the social and supernatural significance of human life, the definitive confirmation of the particular judgment of individuals for their salvation or damnation.[8] Evil will be abolished for ever, and its tool, Satan, utterly vanquished.

There will then take place the 'resurrection of the dead', the work of the Christ who is himself 'the resurrection and the life' (John 11:25).

The dead will arise, the just destined for eternal bliss, the others for damnation, which they will have to suffer in their souls and their bodies.[9]

Christian revelation adds that at the moment of Christ's return the end of the world will take place. The world will undergo a transformation. There will be 'new heavens' and a 'new earth' in which righteousness will dwell (2 Pet. 3:12). Then God will 'make all things new' (Acts 3:21; Rev. 21-22).

In relation to the last things in their collective dimension, the big and disturbing question which arises concerns the existence of risen human beings destined to eternal damnation when elsewhere revelation asserts the total and absolute victory of Christ over Satan and evil, and the transformation of the world and the restoration of all things in a state more perfect than before.

The persistence of the questions we have just mentioned led naturally to the development of eschatological doctrine during the nineteenth and twentieth centuries.

2. CURRENT POSITIONS AND QUESTIONS

Faced with the questions listed above, and in the absence of any answers or adequate explanation, Catholic eschatology in the nineteenth century developed in detailed treatises written in neo-scholastic language which on the whole neither opened up any new directions nor contributed to a deeper understanding of the existential meaning of divine revelation on the ultimate end of man and the universe.[10]

During the twentieth century the new ideas came mainly from the Protestant side. On the subject of the Lord's 'parousia', first of all, Protestant theology based on the form criticism school of exegesis declared bluntly that this had not taken place as Jesus had proclaimed. The liberal theologians concluded that this fact must influence our image of Christ and imply a whole conception of ecclesiology.[11] In contrast, Karl Barth's dialectical theology insisted that 'if Christianity be not altogether restless eschatology, there remains in it no relationship whatever to Christ.'[12]

Bultmann, applying the principle of demythologisation, went even further. Disregarding any relation there might be between specific items of biblical eschatology and events at the end of human life and of history, he stressed the existential implication of the eschatological proclamation. Jesus confronts each individual inasmuch as the kergyma forces us at every moment of our present life to make a fundamental choice, a decision of faith with regard to Jesus, and, in the moral sphere, with regard to the demands of our salvation.[13]

Among Catholics the reaction to the new movements in Protestant eschatology was neither simple acceptance nor systematic opposition. Until recently there were few notable developments. Hans Urs von Balthasar summed the situation up aptly when he divided Catholic production on eschatology into four groups: general surveys which were produced as though nothing had happened for fifty years, an abundance of *haute vulgarisation*, individual studies as yet uncoordinated, and finally hardly more than one or two attempts at a comprehensive eschatology representative of our time.[14] Or, as T. Rast put it, 'There is practically no attempt at a complete eschatology representative of our time'.[15]

Here we shall simply note, following Rast, the main general tendencies of this new eschatological thought.[16] We single out five:

(a) *From eschatology as the last chapter of dogmatics to the essential eschatological situation of all theology*.[17] It is admitted today that a dogmatics based on the history of salvation is determined and characterised by the perspective of the 'eschatological moment'. Moreover, this perspective is involved in the theological appreciation of any Christian doctrine, as Vatican I clearly recognised: 'Reason enlightened by faith, when it assiduously, devoutly and soberly seeks may, with God's help, attain a measure of knowledge of the mysteries, which may indeed be very fruitful, both by analogy with what it discovers in the natural realm and from the connections of the mysteries among themselves *and with the last end of man*.'

(b) *From a future eschatology to eschatological utterances for the present.*[18] The future of humankind already exists in the present. We have a keener awareness today that our final and definitive destiny is even now being shaped and fixed day by day.

(c) *From an eschatology of essence to an eschatology of existence.*[19] Since eschatology is already implicit in the present, Christians are involved in their destiny more existentially.

(d) *From the 'eschata' as places or things to the Eschaton in person.*[20] Consideration shifts directly to Christ, who is the eschatological real presence, though already actual in the history of the Church and the world. God is the 'last end' of creation.

(e) *From individual to universal eschatology.*[21] Eschatology today takes a greater interest in the fate and final destiny of the history of humankind, of the Church and of creation as a whole.

This last current includes the new perspectives of process theology and those deriving from the vision of Teilhard de Chardin.

3. NEW PERSPECTIVES

The currents in current eschatological studies mentioned above are incorporated into the new perspectives by being placed in the context of contemporary science, notably the cosmological and human sciences.

First, the biological sciences have moved from the fixed 'creationist' view of the world to the dynamic 'evolutionist' view. The evolutionary principle is also influencing the world of the physico-chemical sciences, psychology and the social sciences. Similarly the theory of relativity and quantum physics have reinforced the relativist view in considerations of nature and the interrelation of beings.

Secondly, a unified conception of the universe has also become established today and continuity and solidarity are accepted all the way from the basic atomic elements constituting matter to the living creature, with its culmination in the human person. This scientific transportation must inevitably influence religious ideas.

Thirdly, we must also include the influence of modern philosophical and social ideas, the philosophies of the autonomy of human consciousness, existentialism and personalism; socio-economic theories which encourage work for social progress and individual development through individual initiative and self reliance, if necessary without waiting for the intervention of a God or some other-worldly being.

These new attitudes are incorporated in the doctrinal essays of process theology and Teilhardian thought.

(a) Process theology and eschatology

The theological movement known as 'process theology', which originated in the United States and attracted a following also in Britain, developed from the principles of the 'panphysical' philosophy of Alfred North Whitehead.[22] This philosophy, a unified vision of the universe, is indeed well suited to theological development, and such development is taking place, in an attempt to deepen the meaning and widen the reach of the different Christian mysteries in a new perspective.[23]

Whitehead, who firmly integrated 'God' and religion into his philosophical system, did more than any other scientist to justify and demand the establishment and cultivation of a fruitful relationship of mutual support between scientific knowledge and religious values.

Science and religion, he wrote, are the two most powerful forces which influence human beings, though there appears to be a contradiction between the institutions of religion on the one hand and, on the other, meticulous scientific observation and logical deduction.[24]

Briefly, the eschatological perspective of the process theology resulting from Whitehead's philosophy is as follows:

(i) Man, particularly through his body, is physically an integral part of the universe. The environment of the human body includes all human beings and all creatures, even inanimate ones. The destiny of each individual is inseparable from that of all humankind and of the universe. Human life is more than a succession of events between birth and death. There is an assurance that death and the phenomenon of the decay of creatures is not the last word about life.[25]

(ii) The Christian faith, for its part, proclaims that human beings have a twofold destiny, to live this earthly life to the full and to share permanently in God's eternal life. These two destinies are social in nature. To live in love means to lead an existence in society, alive to the needs of others and working to create communities in which justice reigns. To share in God's life is to go through the judgment of God and enter eternity in solidarity with all creation.[26]

(iii) So eschatology is made real in the present life through efforts to establish a 'kingdom of God' lived in communities of fraternal sharing, charity and work for justice. It is not so much at physical death, as in solidarity with other human beings and with the fate of all creation that we shall be judged and admitted to a share in the heavenly kingdom of God.[27]

In this way process theology, although still only at the beginning of its systematic development, seems to have a lot to offer to the discussion of eschatology, as to that of the other Christian mysteries.[28]

(b) The Teilhardian eschatological perspective

The scope of this article does not allow for a lengthy discussion of Teilhard de Chardin's cosmological and anthropological vision, with its connection with the final destiny of man and all creation. There is a full study of Teilhard's eschatology in the extensive thesis of G.-H. Baudry, *L'Eschatologie dans l'oeuvre de Teilhard de Chardin.*[29]

I shall pick out the main lines of the Teilhardian view.

Teilhard de Chardin's thought is above all teleological. Gérard Baudry distinguishes in it two levels with regard to eschatology, that of natural eschatology and that of Christian eschatology.

(i) At the level of natural observation Teilhard de Chardin finds the most striking fact of human evolution to be its 'convergence' towards unity, the 'continual growth of social unification' through human planetisation, the 'growth of generalised technology and industrialisation', the growth of a shared vision of the world which results from this.[30]

(ii) Teilhard believes that humankind, having converged in this way, will reach the normal term of evolution more by a special 'emergence' than by exhaustion or senescence. He sees humanity and the world merge in a magnificent final metamorphosis. 'In this perspective humanity, attached to the end to its planetary support, would end its existence there or, more precisely, would detach itself from it, not as the result of any external disaster or of any internal disease or exhaustion, but by reaching a certain critical state, beyond which we will be able to distinguish nothing more in the future—for the simple reason that this will be a genuine "critical point" of emergence, as it were of emersion, out of the spatio-temporal matrix of the universe.'[31]

(iii) At the theological level Teilhard's eschatology is based on his Christocentric vision. ' "Everything in the cosmos is for Spirit"; that, in natural terms, is the verse "Everything, in Spirit, is for Christ": and that, in supernatural terms is the verse of the Gospel that our modern world needs.'[32] The Omega-Christ is the point in which all evolution culminates. Teilhard also calls Christ 'cosmic', 'universal', 'total', 'the evolver', 'the consummator'.

(iv) The Church, which is, at its deepest level, the 'mystical body of Christ', is the place and the organisation which shapes humankind for eschatology and propels man towards it. The Church is no less than the great 'eschatological axis of humanity'. The Church, Teilhard writes, 'the consciously Christified portion of the world—the Church, the main seedbed of inter-human affinities through super-charity—the Church, the central axis of universal convergence and the exact point at which the encounter between the universe and the Omega point bursts into life.'[33]

(v) The parousia, the return of the Lord at the end of time, is envisaged by Teilhard as marking the end of evolution and bringing about the last stage of hominisation in a supreme transformation. 'In the Christian world's heaven the parousia (or return of Christ at the end of time) occupies a central place even though, as centuries of waiting have passed, it is easily forgotten. In this unique and supreme event in which history (as faith tells us) is to be welded to the transcendent, the mystery of the incarnation reaches its climax and is vindicated with the realism of a physical explanation of the universe.'[34]

In short, as Baudry concludes, the parousia for Teilhard is the manifestation and final guarantee of the full success of hominisation. In contrast to the traditional view, which saw it as the event which was to intervene after a catastrophic end to the history of the world, the parousia will crown in beauty the upward march of humankind towards its full perfection.

'. . . Some people, I know, tend—and delight in doing so—to place history between two 'catastrophes', the accident of original sin and the accident of the parousia. I prefer the alternative vision of hominisation in between two critical points of reflection (the individual reflection at the beginning and the final planetary reflection). This is the only possible vision in terms of 'cosmogenesis' (the 'catastrophe' solution being the solution in terms of a static cosmos). It is also, in my view, the only vision capable of providing a basis for the Christianity of tomorrow.'[35]

(vi) Positively, Teilhard regards the eschatological perspective more as a stimulus to action on earth to prepare for the coming of the metamorphosis of creation than as an object of pure contemplation while we anxiously wait for the final approval of the balance of man's moral performance. 'Some say, "Let us wait patiently until the Christ returns". Others say, "Let us rather finish building the earth". Still others think "To speed the parousia, let us complete the making of man on earth".'[36] The last is Teilhard's own position.

(vii) Teilhard's eschatological vision is so determinedly optimistic that questions have been asked about the role it allowed to evil and sin, and so to the dogmatic thesis of the existence of hell,[37] where there is an impression of doubt or uncertainty.

This is the main criticism made of Teilhard's eschatological vision.

(viii) This apart, in general Teilhard's vision, which in some ways takes up and extends the work of recent studies in eschatology, is a most important contribution. It is a contribution to forming a general attitude, and in particular reinforces the practical attitude which should dominate thinking about eschatology.

'One of Teilhard's great merits,' G.-H. Baudry quite rightly remarks, 'is to have shown that the eschatological perspective is not, as is often believed, a disincentive to action (a flight from the world), but, on the contrary, a stimulus to greater action. His eschatology does not entail simply an orthodoxy, but also an orthopraxis.'[38]

These, then, very schematically presented, are the main lines along which ideas in eschatology are currently moving. The most important task for the present is, while taking a firm stand on the statements of revelation as elucidated by critical scholarship, to incorporate into the Christian theology of eschatology the views suggested by current cosmological ideas and by a better understanding of anthropological structures.[39]

Translated by Francis McDonagh

Notes

1. General information on modern cosmology is available in R. Ruyer *La Gnose de Princeton* (Paris 1974); see esp. E. A. Milne *Modern Cosmology and the Christian Idea of God* (1952) and J. Merleau-Ponty *Cosmologie du XXe siècle* (1965).
2. See 2 Cor. 4:12.
3. See A. Winkelhofer *Das Kommen seines Reiches* (Frankfurt 1959) p. 309.
4. See Denzinger, 1963 ed., 854ff, 1000, 1304.
5. See Denzinger, 838, 1580, 1820, 3548. On the history of the doctrine of purgatory, see the thesis by J. Ntedika *L'Evocation de l'au-delà dans la prière pour les morts* (Louvain and Paris 1971).
6. See Denzinger 1011.
7. See Denzinger 44, 76, 150, 852, 859.
8. See Denzinger 72, 76, 340-342, etc.
9. 1 Cor. 15; 1 Thess. 4:13ff.
10. See P. Muller-Goldkuhle *Die Eschatologie in der Dogmatik des 19. Jahrhunderts* (Essen 1966).
11. A. Loisy wrote: 'Jesus proclaimed the kingdom and the result was the church' (See *L'Evangile et l'église*, 1902).
12. Karl Barth *The Epistle to the Romans* (London 1933) p. 314.
13. R. Bultmann *History and Eschatology* (Edinburgh 1967).
14. See T. Rast 'L'Eschatologie' in *Bilan de la Théologie du XXe siècle*, 2, pp. 501ff. Von Balthasar is quoted on pp. 507-508.
15. T. Rast, in the chapter cited in note 14, p. 508.
16. In the chapter mentioned, T. Rast has a good summary of contemporary movements.
17. *Ibid.* pp. 509-512.
18. *Ibid.* pp. 513-515.
19. *Ibid.* pp. 513-515.
20. *Ibid.* p. 515.
21. *Ibid.* p. 516.
22. See G. Helal *La Philosophie comme panphysique (la philosophie des sciences de A. N. Whitehead)*, (Montreal 1979); A. Parmentier *La Philosophie de Whitehead et le problème de Dieu* (Paris 1968).
23. See J. B. Cobb and D. R. Griffin *Process Theology. An Introductory Exposition* (Philadelphia n.d.); *Process Theology. Basic Writers by the Key Thinkers of a Major Modern Movement*, ed. E. H. Cousins (New York and Toronto 1971).
24. A. N. Whitehead *Science and the Modern World* (New York 1967) pp. 181-182.
25. See Cobb and Griffin, the work cited in note 23, pp. 115-116.
26. See M. H. Suchoki *God, Christ, Church. A Practical Guide to Process Theology* (New York 1982) p. 163.
27. *Ibid.* pp. 174-175.
28. An indication of the way in which process theology is applied to the different aspects of the Christian mystery may be found in Cobb and Griffin, in the work cited in note 23. For the essay on eschatology, see pp. 111-127.
29. G.-H. Baudry *L'Eschatologie dans l'oeuvre de Teilhard de Chardin*, (thesis), (Paris 1977);

see also his article 'Les grands axes de l'eschatologie teilhardienne, 1946-1955' in *Mélanges de Science Religieuse* 34 (1977) 213-235; 35 (1978) 37-71.

30. See Baudry, in the article cited in note 29, 216.

31. Teilhard de Chardin 'Trois choses que je vois' (1948), in *Oeuvres* (Paris 1962) XI p. 168.

32. Teilhard 'Note on the Presentation of the Gospel in a New Age' in *The Heart of Matter* (London and New York 1978).

33. Baudry, in the article cited in note 29, 45ff.

34. *'Trois choses que je vois'*, 168.

35. Quoted by Baudry, in the article cited in note 29, 58-59.

36. 'The heart of the problem', *The Future of Man* (London and New York 1964) p. 260.

37. See *Le Milieu divin* (London and New York 1960) pp. 140ff.

38. Baudry, in the article cited in note 29, 70.

39. Some light on issues such as the judgment after death, the reason for and the nature of hell and the state of the soul before the resurrection might come from the religious experiences of ancient Egypt, Black Africa, Buddhism and Hinduism. See H. Bonnet *Reallexikon der ägyptischen Religionsgeschichte* (Berlin 1952); T. Fourche and H. Morlighem *Une Bible noire* (Brussels 1973); L. V. Thomas *Cinq essais sur la mort africaine* (Dakar 1968); Sri Aurobindo *Birth Centenary Library* 16 vols (Pondicherry 1972).

PART II

Topical

James Buchanan

Creation and Cosmos:
The Symbolics of Proclamation
and Participation

GIVEN THE present situation, by which is meant the general condition we call modernity, no topic could be more timely yet more difficult to reflect upon than is that of 'cosmology'. The need to reconsider cosmology, in terms of both its modern constructs and its archaic structures, is as obvious as the threat posed by modernity to our participation within that cosmos. It is appropriate and essential that any sustained consideration of cosmology commence with some reflections upon the notion of creation and its relationship to cosmology. To that end this article will focus upon the creation narratives found in the book of Genesis in terms of certain interpretive trajectories which have persisted within the Judaeo-Christian tradition.

1. TRAJECTORIES: MANIFESTATION AND PROCLAMATION

The term 'trajectories' is being used here to refer to traditions of interpretation[1] which exist within the broader Judaeo-Christian tradition. In describing manifestation and proclamation as trajectories, the claim is not that these represent two mutually exclusive traditions of interpretation. In fact this article will argue against such a view. On the other hand, their description as such is useful in sorting out two general, yet central orientations within the broader tradition of interpretation.

The designation, manifestation and proclamation, is drawn from the work of Paul Ricoeur.[2] Ricoeur's own use of the terms concentrates its analysis upon the distinction between verbal and preverbal religious expressions in order to show how the verbal orientation of Judaeo-Christianity is 'destructive of the sacred'.[3]

It is in the work of Mircea Eliade that the fullest understanding of religious experience and expression as manifestation is found.[4] In Eliade's understanding of 'archaic ontology' is to be found the key to the manifestation trajectory. According to Eliade, archaic *homo religiosus* experienced himself in terms of a radical participation within a sacred cosmos. In this mode of being, the numinous finds expression in terms of the cosmos itself. Cosmos, in both its spatial and temporal dimensions, participates within its own sacred presence. This presence being the self-affirming experience of participation within the whole, the saturating presence 'manifests' itself spatially as 'hierophany', as rock, tree, etc., and temporally as 'illad tempus', as the time of origin. While Ricoeur's characterisation of this trajectory as preverbal is essentially correct, the distinction does not apply only to preliterate cultures. Rather, the verbal is also taken as a manifestation of the sacred presence. Religious expression in the form of mythos never sees the verbal as merely conceptual, but actualises the word through its participation within the ritual affirmation of sacred space-time. It is the co-ordination of sacred space, as hierophany and sacred time, as mythos, within the ritual which actualises each in terms of the other. The ritual is characterised by its actualisation of the mode of being of participation.[5]

The proclamation trajectory is characterised by its 'accent upon speech and writing and, generally, upon the Word of God'.[6] Proclamation trajectories, found especially in Judaism, Christianity, and Islam, place tradition and prophecy at their centre and organise their theology around certain fundamental discourses. The aesthetic quality of the manifestation tradition is replaced by the ethical, as the commandment replaces the idol, and as a theology of Name replaces the hierophany. Likewise, the temporal as manifest presence, with its cyclical return to origins, is replaced by a fundamentally historical vision of reality. This represents a change in the nature of the correlation of sacred space and time within ritual, as now 'a historical vector runs through the time of repetition and reactualisation'.[7]

The essential distinction between manifestation and proclamation lies in the nature of participation. This is not to say that the proclamation trajectories are nonparticipatory,[8] for there is also here a sense of radical participation. However, within a proclamation mode of being, there is the potential of giving priority to word over world. Thus, the sense of radical participation is now within reality conceived as a reality of words: as history and traditions; as the results of interpretation and reinterpretation. The consequences of this shift for the notion of cosmos will be discussed below.

2. CREATION: A TRADITION OF INTERPRETATION

This section will take up the notion of creation by focusing upon its interpretation within a representative tradition of Old Testament scholarship. It will concentrate upon the work of two scholars, Gerhard Von Rad and Bruce Vawter.[9] Both scholars are representative of mainline traditions of interpretation within Catholicism and Protestantism, thus providing adequate testing grounds for our criteria of manifestation and proclamation.

Gerhard Von Rad has been one of the most influential of the Old Testament hermeneuts in Europe as well as the United States. In order to understand his interpretation of creation, it must be contextualised within his general theology of the Old Testament. Von Rad's theological approach emphasises the unity of the first six books of the Old Testament. This unity is built around a central motif which focuses the hexateuch upon those discourses which narrate a *Heilsgeschichte* or salvation history for Israel. This primary theological thrust of the hexateuch is drawn from the recitation of certain historical events in which Yahweh intervenes in history to save Israel. These interventions are taken as part of a covenant theology which interprets each new historical episode as the renewal and advancement of the covenant between Yahweh and Israel. In the covenant relationship Yahweh is the active partner while Israel is the recipient. Israel's covenantal duty is obedience to the commandments. The nature of the covenantal relationship between Yahweh and Israel places continued belief and faith as the source of assurance.[10]

It is into this general theology of the Old Testament that Von Rad places his interpretation of creation. He emphasises that the doctrine of creation is not a central subject of Old Testament faith. Von Rad contends that the doctrine of creation is of relatively late emergence because 'it took Israel a fairly long time to bring the older beliefs which she actually already possessed about it into proper theological relationship with the tradition which was her very own, that is, with what she believed about the saving acts done by Yahweh in history'.[11] And further, 'Israel had no divine sustenance, blessing, and protection from an environment that was conceived in terms of myth: what had been opened up for her through Yahweh's revelation was the realm of history, and it was in the light of this as starting-point that the term creation had first to be defined'.[12] Creation in this interpretation is subordinated to and incorporated within the notion of a saving history. For Von Rad, this was the master stroke of Israel. Von Rad sees this as true for both of the documentary sources of the creation accounts (the Deutero-Isaiah and Priestly documents). Creation is interpreted as the starting-point of the *Heilsgeschichte*, and as such is part of the aetiology of Israel. It is thus placed *within* a history and functions as part of a faith that is aimed at salvation.

Bruce Vawter's interpretation follows a similar line. Vawter speaks of a 'history of events' in which the essential thrust is Yahweh's redemptive intervention. Creation is the beginning of this history of events, and as such must be considered as one of the events. As he says, 'Creation and fall are not in Genesis simply theologoumena; they are events like all the other events which succeed them'.[13] Vawter sees in this a universal history of and for Everyman. In doing so, Vawter wants to emphasise that the narratives lend themselves to Christian interpretation.

Vawter does this by placing emphasis upon the Word as central to the creation narratives. For creation, 'word is power; the same idea has penetrated the New Testament and is at the base of the Christian conviction of the efficacy of the word in preaching and sacrement'.[14] Vawter also gives a great deal of attention to the special role assigned to man in the cosmos. In Vawter's interpretation, this is one of the major points that distinguishes the Genesis creation accounts from other cosmogonic narratives of the period. All of these differences, in fact, 'are due to the superior vision of man and his place in the universe'.[15]

It is easy to see, then, that both Von Rad and Vawter interpret creation in such a way as to support Ricoeur's claim that the Hebraic faith is a proclamation tradition. The emphasis upon the word, the linear view of a history of events as *Heilsgeschichte*, and the superior vision of man all *shift the focus of the narrative from a concern with sacred cosmos to one of concern for the salvation of humanity*.

3. CREATION: THE SYMBOLICS OF PARTICIPATION

This section will argue that it is by viewing creation in terms of a symbolics of participation that we can best understand the nature of cosmos within both the manifestation and proclamation trajectories. To develop a theory of symbolism in this limited space is impossible; therefore, the discussion will be limited to a few essential points. These are drawn from the work of Ricoeur, Rahner and Eliade that, in turn, draws upon a long and rich tradition of reflection. The focus of the discussion will be upon the dialectic of experience and expression. This dialectic recognises the inherent mutual attraction and repulsion of the poles. This means that although experience is always unfulfilled in its expression, the self-fulfilment of the experience/expression dialectic is inherent to it as telos. Rahner speaks of this in terms of 'a plurality of being seeking to realise itself as a unity of being through expression'.[16] For Ricoeur it translates into his project of a 'poetics of the will'.[17] Given the dual limitation of expression and pure experience (the Kantian in-itself), we are left to a reality of expression as limit-expression. Thus, we exist within the unfulfilled dialect of experience/expression.

The paradox remains that language carries its own fullness as potential. The compulsion of language, be it scientific, commonsensical, or symbolic, is towards its own fullness. The most radical response of language to its potential of universal content is as symbol. The symbol universalises expression as limit by means of its double intentional structure. It has connection to its appearance as event or object, but it also has connection to a universal content. The symbol, in its striving for universal content 'constitutes the symbolised as its own self-realisation'.[18] As self-realising, the symbol not only is the most radical attempt of expression to recollect experience, but also 'understands' its own double intentionality, meaning that the symbol inherently realises itself in terms of its own limitations. Thus, we can speak of symbols as the self-realisation of the experience/expression dialectic as both 'limit-to' (its inherent particularity) and as 'ground-of' (its potential universality).[19]

The question at hand can be stated: What is the relationship of the limit-experience best symbolised as creation to the notion of cosmos?

The terms 'Cosmology' or 'Cosmos' refer to a family of concepts which consider the world or universe holistically. The first distinction that must be made is between notions of sacred cosmos and the scientific cosmologies. The best descriptions of sacred cosmos can be found in the work of Mircea Eliade (and of course those myths he studies). The sacred universe is based upon sets of correspondences (e.g., macrocosm and microcosm, earth and sky to female and male, between the *in illo tempore* and the order of nature, etc.) while the scientific notion of cosmos describes through the reduction of various relationships to laws. For the scientific notion, cosmos exists within a conceptual space. Phenomenologically speaking, there is a radical difference in the experiential dimension which characterises these two approaches. The key to understanding this difference is to be found in the idea of participation. While the sacred cosmos is marked by a radical participation within it, the scientific emphasises a detached, objective mode which has as its goal a radical non-participation. This change from participation towards non-participation can best be analysed in terms of a perspectival shift described from cosmocentrism to an anthropocentrism, a shift from the view that man exists within cosmic space to one in which cosmos exists within a conceptual space.[20]

If we attempt to relate the two perspectives on cosmos to manifestation and proclamation, it is readily seen that the notion of sacred cosmos relates to manifestation. However, it would be going too far to suggest a pure relationship between the scientific perspective and the proclamation tradition. Rather, it should be said that within the proclamation tradition there is the potential or seed of the techno-scientific approach. It is in terms of creation that we can best see both the potential of scientism[21] as well as the religious potential within the proclamation tradition. In our description of that tradition, we noted the emphasis upon the verbal that easily gives way to a conceptualist or intellectualist mode of understanding. The danger here should be apparent, and creation is precisely what is at stake. The scientific conception of the cosmos has moved through various stages. The move from the geocentric view of a closed, finite universe to a heliocentric view which sees the universe as infinite also entails the view of the potential infinity of the human mind. The potential infinity of the human mind as a potential conceptual infinity details a radical anthropocentric shift. 'Just as God is the creative centre of all that is so man is the centre and creator of an infinite conceptual region or linguistic space which has room for the cosmos and for God. By forcing reality to assume its place within the borders of that region, man recreates God's creation in his own image.'[22] Thus, the perspective shift is really grounded in the reworking of the notion of creation. But creation also functions as a symbolic corrective to this tendency within the proclamation tradition. Earlier we spoke of the distinction in kind and degree of participation characterising cosmos as sacred and the scientistic view. While the scientistic perspective is a potential, even a tendency, within the proclamation trajectory, there is also the fact that the word as event has an experimental dimension. Thus, the experience of language also entails an experience/expression dialectic. We spoke of the symbolic as the result of the eternal urge within language towards its own fullness. There is only one symbol which truly captures the idea and feeling of the fullness of language. This symbol is creation. Only the word as creative actualises its potential. It is only in creation that experience/expression are unified and need no longer exist in dialectical tension. Relative to the fullness of language, creation symbolises the limit-experience of language itself.

This limit-experience of language itself opens us to the possibility of the manifestation orientation. To experience the limit is to realise the limits of conceptual space itself. To do so is to alter ones mode of being from one of detachment to one of participation, that feeling of participation and even dependence essential to the manifestation tradition. Within the manifestation orientation, creation becomes the symbol *par excellence* for this limit-experience of participation. The manifestation mode necessarily finds difficulty with any radical form of anthropocentrism. The view of salvation inherent to this position is that the salvation of man is the salvation of the entire cosmos and thus of man as a participant. This accounts for the emphasis upon cosmic renewal in the manifestation traditions studied by Eliade.[23] It should also be noted that the cosmogonic myth is the central myth within these traditions.

It has been argued that creation is the symbol which best captures the most fundamental limit-experience within each of the traditions. Creation symbolises the self-realisation of limit, linguistically and existentially. Furthermore, creation becomes the central symbol which assures that proclamation and manifestation always exist in a balanced dialectic. It assures this balance to the extent that it functions symbolically, determining that the potential of universal content is experienced as limit. To function symbolically demands that a balance be maintained within the double intentional structure. To lose sight of the particular intentionality in light of the universal is thus to fall into a conceptual trap like that described by Harries.[24] We can say, then, that to the degree that creation functions as a true symbol, proclamation and manifestation remain in a balanced dialectic.

In conclusion, there are two points concerning creation vital for the contemporary discussion. First, those interpretations of creation cited in the second section of this article understate the case for the existence of a manifestation tradition within the Israelite religion. Scholars such as Claus Westermann, Robert Cohn and even Eliade find evidence within the scriptures of a manifestation orientation. Cohn has produced an interesting study of the notion of sacred space within the Old Testament.[25] Cohn does not see any necessary connection between the idea of a *Heilsgeschichte* and the desacralisation of nature. His study demonstrates a strong sense of sacred cosmos and participation in the Hebrew scriptures. Westermann implies the same conclusion in his interpretations of creation.[26] Westermann places creation and redemption in tension. 'It is essential for understanding the Old Testament that the relationship between creation and redemption consists in polarity.'[27] He sees the 'historical credo' found in Von Rad and Vawter as the result of the incorporation of notions of creation within the all-embracing ideas of faith and revelation. According to Westermann, such interpretations are not true to the texts, opting to interpret the Old Testament from the perspective of the New Testament where there is more justification. Westermann contends that the Old Testament does not propose creation as either an article of faith or as a product of revelation but understands creation existentially. Thus, the creation narratives are to be considered myth and not placed within the idea of history. He agrees with Eliade that opposing myth to history is to misunderstand myth altogether. As he says, '. . . myth belonged originally to the context of survival, an expression therefore of one's understanding of existence'.[28] This interpretation is also strengthened by the fact that there has remained within the Jewish tradition a vital strain of manifestation orientation. One needs only mention Kabbalah and Hasidism to support this claim.[29] The inevitable conclusion is that the interpretations of Ricoeur, Von Rad and Vawter are indicative of a history of effects of interpretations within a Reformed and Catholic hermeneutic more than within the scriptures' own historical period.[30]

Finally, I would like to suggest that creation is the very heart of Ricoeur's 'logic of meaning in the sacred universe'.[31] Ricoeur has described this logic in terms of the correspondences noted earlier. Granting this, we are not dealing with a formal logic here but, rather, with a transcendental logic. It is a logic of experience, even a *'logique du coeur'*.[32] It is the radicalised experience of participation best symbolised as creation which is at the heart of the correspondences. In an age in which our relationship to the cosmos is one of almost wanton destruction and in which we are increasingly confronted with the fact that this wanton destruction is nothing short of self-destruction, the centrality of creation as the symbol of that limit-experience of utter participation within the cosmos has never been more crucial. The vital importance of confronting this experience and taking it to the heart of one's existence is now, as it has always been, a matter of survival. As *homo religiosus* the demand of creation is the experience of ourselves as creatures among creatures, as full participants within a cosmos whose desacralisation can mean only self-desacralisation.

Notes

1. 'Traditions of interpretation' is equivalent to 'history-of-effects'. See H.-G. Gadamer *Truth and Method* (New York 1975) pp. 267-274, 305-345.

2. Paul Ricoeur 'Manifestation and Proclamation' *Journal of the Blaisdell Institute* (Winter 1978) 13-35.

3. *Ibid.* 13.

4. Particularly *The Myth of the Eternal Return or Cosmos and History* (Princeton 1974); *Patterns in Comparative Religion* (London 1958); *The Sacred and the Profane: The Nature of Religion* (New York 1961).

5. For the best theological discussion of manifestation and proclamation, see David Tracy *The Analogical Imagination: Christian Theology and the Age of Pluralism* (New York 1981) pp. 193-229.

6. Ricoeur, in the article cited in note 2, p. 21.

7. *Ibid.* 22.

8. This is Tracy's characterisation, but his use of 'participation' is not the same at this point.

9. Gerhard Von Rad *Genesis* (Philadelphia 1961) and *Old Testament Theology* (New York 1962). Bruce Vawter *On Genesis: a new reading* (New York 1977).

10. Von Rad *OTT* pp. 105-135.

11. *Ibid.* p. 136.

12. *Ibid.*

13. Vawter, in the work cited in note 9, p. 31.

14. *Ibid.* p. 41.

15. *Ibid.* p. 52.

16. Karl Rahner *Theological Investigations* IV 'The Theology of Symbol' (Baltimore 1966) p. 234.

17. See particularly *Freedom and Nature: The Voluntary and the Involuntary* (Evanston 1966) and *The Symbolism of Evil* (New York 1967). See also Ricoeur 'The Hermeneutics of Symbols and Philosophical Reflection' in *The Philosophy of Paul Ricoeur* (Boston 1978) pp. 36-58.

18. Rahner, in the article cited in note 16, p. 242.

19. See David Tracy *Blessed Rage for Order* (New York 1975).

20. For a good analysis of the 'perspectival shift' see A. Koyre *From the Closed World to the Infinite Universe* (Baltimore 1957); also M. Muntz *Theories of the Universe from Babylonian Myth to Modern Science* (Glencoe 1957); J. Merleau-Ponty and Bruno Morando *The Rebirth of Cosmology* (New York 1976).

21. Scientism is best conceived as the combination of science and technology. It is characterised by a linear thinking whose intention is manifest as will-to-power. See Heidegger 'The Question Concerning Technology' in *Basic Writings* (New York 1977).

22. Karsten Harries 'The Infinite Sphere: Comments on the History of a Metaphor' *Journal of the History of Philosophy* 13 (Jan. 1975) 13.

23. Eliade *The Myth of the Eternal Return* (Princeton 1974).

24. Harries, see note 22.

25. Robert Cohn 'The Shape of Sacred Space' *Four Biblical Studies*, (California 1981).

26. Claus Westermann *Creation* (Philadelphia 1974) and *A Thousand Years and a Day* (Philadelphia 1962).

27. *Ibid*. p. 117.

28. *Ibid*. p. 12.

29. Being uninformed concerning the Jewish scholarship I wish to thank Susan Shapiro for this insight. See Gersham Scholem *Major Trends in Jewish Mysticism* (New York 1977), Martin Buber *I and Thou* (New York 1958).

30. Ricoeur has attenuated this position somewhat in his more recent writings. See particularly his 'Toward a Hermeneutic of the Idea of Revelation' in *Essays on Biblical Interpretation* (Philadelphia 1980), pp. 73-119.

31. Ricoeur uses this phrase to sum up the manifestation tradition; see: 'Manifestation and Proclamation' pp. 19-21.

32. Pascal's famous phrase.

Hermann Brück

Astrophysical Cosmology

SCIENTIFIC COSMOLOGY is concerned with the study of the structure and evolution of the physical Universe as a whole. When we consider that we are dealing here with huge dimensions both in space and time, cosmology appears to have a near impossible task. That it has been pursued with considerable success is due to the recognition that in its large-scale structure the Universe is relatively simple and also to the fact that the finite velocity of light allows us to follow the evolution of the Universe back into its past by observing more and more distant parts of it.

Modern scientific cosmology came into being in the 1920s. It started on the observational side with the discovery of the nature of 'galaxies' which are the building blocks which define the large-scale structure of the Universe. On the theoretical side cosmology began with the application of Einstein's theory of general relativity to various theoretical models of the Universe.

Astronomical observations tell us that the Universe is made up of many millions of galaxies which are scattered fairly evenly through space and which themselves are structures composed of millions of individual stars and some diffuse interstellar matter. 'Our own galaxy', the galaxy in which our Sun is placed, contains some 100,000 million stars, all more or less similar to the Sun. The distances between the stars are of the order of several light years, one light year being the distance light travels in a year, about 9½ million million kilometres. The diameter of our galaxy is about 100,000 light years.

Large as the distances between stars are, the distances between galaxies are about a million times greater. As the American astronomer E. Hubble found in the early 1920s even 'nearby' galaxies are several million light years away from us, and the distances of very faint galaxies amount to several thousand million light years. In observing them we see them not as they are now, but as they were several thousand million years ago.

A second major discovery which the same American astronomer made in 1929 concerned the motions of the galaxies. Observing their spectra and using the well-known Doppler principle, Hubble measured the motions of galaxies in the line of sight and found that they move away from us with speeds of recession which increase with their distances. Such motions indicate that the galaxies also move away from each other or that that the whole realm of galaxies is in a state of general expansion. Observations show that this expansion proceeds at a fast rate: the most distant observable objects recede from us and from each other with speeds which are substantial fractions of the speed of light.

44

The discovery of this general expansion suggests that the present condition of the Universe is the result of an explosion in the past of what has been called a 'primeval atom' which contained in a state of extreme compression all the material which we now see scattered through the Universe. If the presently observed rate of expansion has been more or less the same from the beginning to the present time—and this is a not unreasonable assumption—then the original explosion, the 'Big Bang' as it is generally called, must have occurred some 15,000 million years ago.

Moving from observation to theory, there have been several pioneers in the field apart from Einstein, but the 'Father of Big Bang Cosmology' was the late Monsignor George Lemaitre of Louvain. He was one of the earliest members of the Pontifical Academy of Sciences and its President from 1960 until his death in 1967. It was in his honour that the Academy held in the autumn of 1981 a Study Week on 'Cosmology and Fundamental Physics' in which about twenty of the world's leading astronomers and physicists took part.

The inclusion of fundamental physics in the subject of that Study Week recognises the fact that the earliest stages of the exploding Universe were also those which produced the fundamental elementary particles of physics out of which all others and indeed all structures of the present Universe evolved in time.

When we speak of the 'beginning' of the Universe, we always mean the time to which we can trace back ultimately all the various phenomena which we now observe. Though physical conditions in the earliest stages of the Universe were very different from those with which we are familiar now, modern nuclear physics is fully capable of envisaging those early conditions and is indeed able to draw reasonable conclusions about the state of the Universe only minutes, seconds or even fractions of a second after the Big Bang.

Radio astronomy from its earliest days has played a major role in the study of the history of the Universe. Starting with the pioneering work of Sir Martin Ryle in Cambridge in the 1950s radio observations of the distribution in space of distant galaxies demonstrated conclusively that in its past history the Universe was decidely different from what it is now and that it shows every sign of cosmic evolution. At the Vatican Study Week particular attention was paid to the distribution of quasars; these are extremely luminous radio galaxies which are detectable to very great distances thus giving important information about early periods in the history of the Universe.

Again in the field of radio astronomy a major discovery was made in 1965 by two American physicists, A. A. Penzias and R. W. Wilson, who, using a radio telescope in the microwave range, detected the existence of a weak radio radiation which can be best interpreted as a relic of the radiation which was emitted by the Universe when it was a thousand times smaller and hotter than it is at present. The energy of this microwave radiation corresponds to a present temperature of the Universe at large of only three degrees above the absolute zero, the temperature to which the radiation has cooled as a result of expansion since it was emitted about a million years after the Big Bang.

Compared with the 15,000 million years which have elapsed since the beginning of the expansion an 'age' of a million years marks still a very youthful stage in the history of the Universe. However, as has been mentioned earlier, the condition of the Universe can in fact be traced back to much earlier times by the laws of fundamental physics. Such extrapolations become naturally more and more tentative the further one goes into a range where those laws may not be any longer strictly applicable.

What is certain is that the early Universe in its hot compressed state contained a great deal of energy which according to the precepts of quantum theory and general relativity can appear in the form of either radiation or matter. A few seconds after the Big Bang most of the energy of the Universe was actually in the form of radiation, but a small fraction of it was in the form of matter. A few minutes later physical conditions were such that the element helium could be produced in addition to hydrogen, the most abundant of the elements.

As regards the heavier chemical elements it had been known since the work in the late 1950s of E. M. and G. R. Burbridge, W. Fowler and F. Hoyle that these are produced in nuclear reactions in the hot interiors of massive stars. The ingenious theory of nucleogenesis could not account, however, for the production of the element helium, and the discovery that this element could be produced in its correct proportion to hydrogen shortly after the Big Bang provided a welcome solution to a long-standing problem. Further confirmation of this result and the explanation of the origin in the Big Bang of other light elements like deuterium and lithium were presented at the Vatican Study Week giving the strongest support to the Hot Big Bang theory of the history of the Universe.

There are, of course, many open questions. One important one concerns the problem of how and at what stage galaxies were able to condense out of the diffuse background matter and how structures consisting of clusters and even superclusters of individual galaxies with large voids in between came into being and evolved with time. The formation of these structures took place at a relatively late stage, perhaps a hundred million years after the Big Bang, and it is an interesting fact that the early period from one second to a million years in the history of the Universe is better understood than the later one of a million to a billion years.

Having considered the past history of the Universe we may well ask about its future and its eventual fate. Will the present expansion go on for ever or will it slow down and will the Universe ultimately collapse? According to relativity theory the rate of expansion of the Universe depends on the amount of matter it contains. If the average concentration of matter is above a certain critical density gravitational forces will brake the expansion and the Universe will ultimately re-contract. It is easy to calculate this critical density, which is found to amount to about three atoms per cubic metre, a very small density indeed! When one compares this with the actually observed density of matter found from averaging out the material which is contained in galaxies, one finds that this density is only about one-thirtieth of the critical one. Matter appears to be very thinly spread and the Universe in the large looks very empty.

However, investigations of the relative motions of members of clusters of galaxies point to the existence of a considerable amount of unseen matter, and at the Study Week it was agreed that galaxies may well contain as much as ten times more dark matter than is actually observed. Such matter might push the average density up to the critical value making the Universe collapse in the far future.

Dark matter could exist in various forms. One possibility which was discussed at the Study Week, is that we are dealing here with neutrinos, elementary particles which were formed in the Big Bang and which, though individually of negligible mass, may because of their large numbers contribute a substantial amount to the material contained in the Universe.

The amount of matter in the Universe controls the rate of its expansion and this again determines the formation and evolution of galaxies and stars. Outside of a fairly narrow range in the rate of the expansion the Universe could not have produced the stars which in their turn formed the heavy chemical elements of which the environment of our planet and ultimately we ourselves have been formed. In a faster expanding Universe the gravitational forces needed to gather matter together would not have been strong enough to compete with the dispersing effect of the expansion. In a more slowly expanding Universe on the other hand the Universe would have halted and re-collapsed before the stars had time to evolve. There are those who find these circumstances more than a coincidence and who have proposed the so-called Anthropic Principle which states that our Universe is unique in the sense that no other could have led finally to the appearance of intelligent life within it. The phrase '*Cogito ergo mundus talis est*' has been coined by the proponents of that principle. The coincidences are certainly remarkable, but cosmologists in general do not find it necessary to adopt this particular view, and the topic did not form any part in the deliberations of the Study Week.

In summary, it is possible to say that all the scientific evidence, both observational and theoretical, favours the Hot Big Bang of the Universe in broad outline. In detail, as we have mentioned, some phases and in particular the earliest ones remain as yet unclear. Though a fraction of a second may be an inappreciable interval in human experience it is not so in the realm of particle physics, and the time to which the behaviour of the Universe can be extrapolated backwards will certainly be pushed further back with the help of new observations and improved understanding of fundamental physics. Cosmologists make no claim that they will ever reach the instant of the actual 'beginning' though in the mathematical models of space and time which S. W. Hawking described at the Study Week it is perfectly possible to consider earlier periods.

Because of occasional misunderstandings it is important to stress that the instant at which the expansion of the Universe began has nothing whatever to do with the theological concept of creation where something comes into existence out of nothing whether it begins in time or not. It was unfortunate that this distinction was not upheld by as distinguished a mathematician as Sir Edmund Whittaker when in his Riddell Lectures of 1942 he said: 'When by purely scientific methods we trace the development of the material Universe backwards in time, we arrive ultimately at a critical state of affairs beyond which the laws of nature, as we know them, cannot have operated: a Creation in fact. Physics and astronomy can lead us through the past to the beginning of things, and show that there must have been a creation.' And again in his Donnellan Lectures of 1946 when he stated: 'Different estimates converge to the conclusion that there was an epoch about 1,000 or 10,000 million years ago, on the further side of which the cosmos, if it existed at all, existed in some form totally different from anything known to us: so that it represents the ultimate limit of science. We may perhaps without impropriety refer to it as the Creation.'

Sir Edmund Whittaker's ideas found a favourable response with Pope Pius XII who in an address to the Pontifical Academy of Sciences in 1951 fully endorsed Whittaker's views when he said: 'Thus with the concreteness which is characteristic of physical proofs, science has confirmed the contingency of the Universe and also the well-founded deduction as to the epoch when the cosmos came forth from the hands of the Creator. Therefore, God exists! Although it is neither explicit nor complete, this is the reply we were awaiting from science.' The pope's words received considerable publicity at the time and the underlying misinterpretation of scientific evidence caused much embarrassment.

As another example of the misunderstanding which can arise from the identification of the physical beginning with the creation of the world we may recall the quite unnecessary controversy which surrounded for some time the 'steady-state cosmology' which was developed in the late 1940s by H. Bondi, T. Gold and F. Hoyle. In this cosmology the Universe is seen on the large scale at least as being always the same and having neither a beginning nor an end. To account for the undisputed phenomenon of the dispersal of the galaxies the authors of the theory introduced the idea of 'continuous creation' according to which new matter comes into being continuously at a rate which makes up for the loss of matter due to the expansion of the Universe. The idea that there was no beginning was favourably received by those who thought that it would do away with the need for a creation. It also appealed to some astronomers at the time because it seemed to circumvent a discrepancy which then existed, but which was soon afterwards explained, between the age of the whole Universe and the ages of some of its component parts. The steady-state theory has had to be abandoned on observational grounds though there was confusion in some quarters regarding its so-called philosophical and theological implications.

The entirely separate questions of the beginning of the physical Universe and the creation of the world and their relation to each other have often been discussed since the days of Whittaker; we may mention in particular E. L. Mascall's *Christian Theology and Natural Science* of 1956 or Stanley L. Jaki's *Cosmos and Creator* of 1980. On the question of a beginning in time Mascall quotes St Thomas in his *Summa Contra Gentiles* where the matter is put very simply: 'God brought into being both the creature and time together' and equally clearly: 'The preservation of things by God does not take place by some new action, but by a prolongation of that action by which He gives existence; and this action is without change or time.'

Mary Hesse

Cosmology as Myth

FROM THE scientific point of view the conflict between Christian doctrines of creation and scientific cosmology has been predicated upon a particular philosophy of science which is at once *realist* with regard to scientific theory, and *objectivist* with regard to truth. Thus, on the one hand scientific cosmology has been taken to be ontological in its implications, and on the other hand the notion of myth has been devalued as the antithesis of fact and truth. Consideration of the conflict from a philosophical point of view therefore requires scrutiny of both these presuppositions.

1. REALISM

Since the inception of modern science in the seventeenth century, it has generally been claimed by scientists and philosophers alike to be the definitive method of obtaining knowledge of reality. In the seventeenth century this view took the form of a dogmatic mechanism—natural reality is nothing but matter in motion, and it is the task of natural knowledge to uncover the corpuscular systems below the level of observation which will explain all the multifarious phenomena observed. The same realism about theoretical discovery informed nineteenth-century claims that 'science has shown' nature to be both mechanist and determinist, subject to the rigid regime of Newton's laws on motion.

Two twentieth-century developments have shown this view to be untenable. The first arises from the refutation in modern physics of both mechanisms and determinism—the basic laws of nature are no longer Newton's but those of relativity and quantum physics. These are more tentative and open to conflicting interpretations than Newton's ever were, and in particular quantum physics has contradicted the Newtonian theory that all physical processes are determined by laws plus the total set of initial conditions at any given time. Secondly, philosophical reflection upon science has clarified the distinction between the claim on the one hand that scientific theory is ideally a realistic and comprehensive description of the natural world, and on the other hand the patent fact that science is in general ever more successful in providing correct expectations about the workings of the natural world. This instrumental success of science is, of course, relatively local, and although realists claim universality for theories we can only test their adequacy in the limited space-time environment in which we can make observations. But within this local environment there is factual constraint upon theory, or, to put it the other way round, what we count as 'factual' is just that which we find to be the objective, value-neutral grounds for successful prediction and control of nature. The 'reality' which instrumental science is concerned with is just this 'objectified' nature which constrains our beliefs and actions relative to our external environment, and the 'truth' which is appropriate to it is truth which 'corresponds' with objectified reality.

It is, of course, this instrumental aspect of science that guarantees its prestige in the public mind. But the realistic claim that science can ideally attain true universal theories of the world does not follow from its instrumental success. Scientific theories are always in fact *undetermined*—there is always a multiplicity of theoretical interpretations that fit the facts well enough.[1] This is nowhere more true than in scientific cosmology, where the proportion, as it were, of fact to theory is relatively small, and where conflicts between fundamental theories must be expected to continue as long as empirical investigation lasts.

At this point it will be useful to give a very simplified classification of the possible relations between the truth-claims of scientific cosmology and of religious doctrines of creation:

(a) Cosmology and doctrine both claim to describe the same objectified reality, but do so in conflicting ways.

(b) Cosmology claims definitive descriptions of reality, but doctrine claims some non-descriptive form of 'truth'.

(c) Cosmology makes no claim to definitive description of reality, but doctrine does.

(d) Neither cosmology nor doctrine claim definitive description of reality. The 'truth' appropriate to both is non-descriptive.

It follows from the above arguments for the non-realistic status of scientific theory that (a) and (b) are untenable. It should be noted of course that even if these arguments fail, conflict might be avoided by adopting alternative (b), and we shall indeed discuss the possibility of non-descriptive forms of religious truth under alternative (d). Alternative (c) lends some comfort to religious fundamentalists, but the philosophical difficulties of defending this sort of claim for religious doctrine are so extreme that it will not be further considered here. It is alternative (d) that is most interesting and important for a modern account of the relations between cosmology and doctrine, and since it so far expressed in merely negative terms, it needs further elaboration of the concept or concepts of 'truth' implied.

2. WHAT IS THE STATUS OF SCIENTIFIC THEORY?

There is no doubt that theory is generally interpreted by both scientists and the general public as a putatively true description of the real natural world. If the foregoing anti-realist arguments are accepted, however, a problem arises about how to understand the increasingly complex theoretical constructions which provide over-whelming fascination for laymen and unending motivation for practising scientists. It would clearly be too simple to ascribe this fascination purely to the instrumental success of science and its technological applications. In view of the profound aesthetic appeal that scientific problem-solving has for those who pursue it, we might be tempted to put its non-instrumental aspects in the same category as 'art' or 'play', which are regarded in positivist philosophy as having no direct social or epistemological function except in their by-products of challenges and pleasure.[2] A great deal of scientific theorising, especially in fundamental physics and cosmology, is not too distant from the creation of science fiction, which might indeed be said to be speculative theory without the full rigour of experimental control.

But the aesthetic appeal of science indicates something deeper than this, of which the claim to 'discover reality' is but a misleading hint. It is quite clear that for the general public in educated Western society, scientific accounts of the origin and destiny of the world, and of the status of human beings within it, have replaced the traditional mythical accounts given in various forms in all religions, including in particular biblical religion. In other words, whatever other significance scientific theory has, it certainly has the status of cosmological myth in our society, as can be seen in the way 'origins' are taught in schools, and in the popularity of media presentations of fundamental science, both of physics and biology.

It would be a mistake, however, to take this mythic function of cosmology as just another by-product of science, or as a one-way interaction between science and society. In the light of our earlier questions about what non-realistic forms of truth might be attributed to science, we should rather say that its mythic functions are of its essence. To explore this idea a little further will help to illuminate the concept of 'myth' both in science and religion.

It has long been recognised in the history of science that factors other than empirical grounds and pragmatic success have entered into the formulation of scientific theory. Alexandre Koyré, for example, in his aptly titled *From the Closed World to the Infinite Universe* (Baltimore 1957), showed how the theory of the heavens developed out of the finite Aristotelian world into the spatio-temporal infinity of Newton's universe, and how this development expressed what Koyré called 'a very radical spiritual revolution' in which man 'lost the very world in which he was living and about which he was thinking, and had to transform and replace not only his fundamental concepts and attributes, but even the very framework of his thought' (*ibid.* Introduction). Similar studies were carried up to the time of Newton by E. A. Burtt in his *Metaphysical Foundations of Modern Science* (London 1924).

Such studies demonstrated the unity of *ideas* in the development of science, and showed that theory is not independent of surrounding theology, ideology and metaphysics. More recently, however, partly under the influence of Marxist interpretations of history, but generally not committed to these, historians of science have begun to explore in detail the social, political and economic concomitants of scientific theory. Where there is in principle a multiplicity of possible theories more or less fitting the facts, and where practical constraints of simplicity and convenience do not reduce this multiplicity to a single solution, many kinds of factor from the surrounding cultural environment may be decisive. It may be thought that examples are easy to find from early science or from the 'soft' non-physical sciences, but hard to find in recent physics. Let us therefore mention two examples of the latter kind which show that extra-scientific influence is not absent even there. In late nineteenth-century Cambridge, Maxwell's electromagnetic theory was interpreted in terms of an aether-substance present throughout space, in contrast to the mathematical action-at-a-distance theory prevalent on the continent. Brian Wynne has argued that acceptance of the aether owed much to a perceived need for antidotes to materialism as a social ideology, and to the fact that many of the practitioners of physics in Cambridge at the time were active in psychical research or sympathetic to such interests. 'Aether' was seen as an intermediary between 'matter' and 'spirit', and thus rendered more acceptable the non-materialist metaphysics espoused by these physicists.[3] Again, Paul Forman has argued that the acceptance of indeterminism in quantum theory in the Germany of the Weimar Republic had roots in the current romantic defeatism and anarchism of post-war culture, as well as in purely empirical arguments.[4]

Both these historical examples are controversial, but the fact that such arguments can be seriously defended in these and other cases indicates that interaction between society and theory is not all one way. Scientific theory, as it is accepted within the scientific community and hence by the wider public, is the result of a complex of decisions and persuasive arguments of individual scientists which reflect their ideological and evaluative commitments as well as their instrumental motivations. Scientific theory has its place in the dynamically interacting mythologies of Western culture just as cosmologies have in primitive cultures, and without input from cultural needs there would be no motivation for instrumental science to flower into the universal world models which have always been part of its essential goal.

3. SCIENCE AND RELIGION AS MYTH

I said earlier that it is a mistake to interpret this pressure towards universalisable science as the search for a comprehensive true theory corresponding to reality, and I have interpreted theory rather as one type of response to cultural needs for myth and ideology. This thesis is but an echo of Durkheim's theory of the social foundations of knowledge, and brings us close also to his social theory of religion.

The theory of religion in Durkheim's classic *Elementary Forms of the Religious Life*[5] may be crudely summarised in the phrase 'society is God'. He argues that in primitive societies religious systems (including myths, beliefs, institutions, ritual actions) are examples of 'collective representations'—they constitute social norms, constraints and obligations into which all members of the society are inducted, and which form an 'objective' part of their environment carrying as much 'reality' in experience as does the natural world. Indeed the distinction between social and natural worlds is not the important one in primitive societies, because the crucial social boundary is drawn rather between the 'sacred' and the 'profane':

> A religion is a unified system of beliefs and practices relative to sacred things, that is to say, things set apart and forbidden—beliefs and practices which unite into one single moral community . . . all those who adhere to them (*ibid.* p. 47).

Whence comes the sanction for the permissions and prohibitions which issue from the realm of the sacred? As a committed positivist with regard to scientific knowledge, Durkheim is unwilling to ascribe sacred power to divinities of whose existence there is no independent evidence, and argues instead that there is an obvious and clearly existing source of such extra-individual constraint, namely society itself. When individuals pray, worship, obey or incur guilt, it is to society as a whole that these religious attitudes are directed. Moreover, these activities have social functions in promoting stability and cohesion without which societies could not survive.

Durkheim's theory is, however, explicitly not a theory which reduces religion to an epiphenomenon entirely by other social structures, as in the Marxist interpretation. Durkheim came to see it rather as a mutually interacting part of the social complex, having a relative autonomy of its own:

> The life thus brought into being even enjoys so great an independence that it sometimes indulges in manifestations with no purpose or utility of any sort, for the mere pleasure of affirming itself (*ibid.* p. 424).

This description parallels nicely the account we have given of the social function of scientific theory, and indeed in Durkheim's thought it is an account which is supposed to have validity not just in regard to religion, but to knowledge systems in general. In an earlier work, *Primitive Classification*,[6] Durkheim and Mauss had laid the foundation of their theory of religion in a much more radical sociology of knowledge. By citing examples from anthropological fieldwork they concluded that all kinds of categories of logic and reason are based on primitive classifications of the natural and social worlds: there are correlations of cosmic spatial and temporal categories with the topological arrangements of the clan's living space and their organisation of intervals of time; notions of natural kinds and natural causality are derived from social classes and social authority; magical laws and rituals involving sympathy, contiguity and participation are symbolisations of social relations; and thence derive ultimately the unified systems of law which constitute science and logic. The theory is explicitly a socialisation of Kantian rationalism. Only the social entity, according to Durkheim, carries the objectivity, impersonality, necessity, autonomy and authority to account for the apparent inescapability and universality of the laws of thought.

Durkheim's attempt to reduce religion to a purely social phenomenon is not, of course, sufficient to *disprove* religious belief in the existence of God, and indeed we shall see in the next section that his own theory of morality needs a faith in something transcending this sociological interpretation. His socialisation of the categories of reason is, however, more cogent, and rests on arguments that are independent of his rejection of the concept of God. Whether right or wrong, this very powerful and far-reaching theory of rationality has not yet been properly assimilated into philosophical thought. In one particular Durkheim himself did not exploit it to the full. *In Implicit Meanings* (London 1975) Mary Douglas points out that there is one crucial area of knowledge that Durkheim excepted from his thesis, namely that of natural science itself. Science provided Durkheim with his one anchor against the shifting sands of relativism, and seemed to him to be the necessary method of grounding any theory of knowledge and society at all. Douglas, however, thinks we should refuse any privileged domain of knowledge:

> . . . anyone who would follow Durkheim must give up the comfort of stable anchorage for his cognitive efforts. His only security lies in the evolution of the cognitive scheme, unashamedly and openly culture-bound, and accepting all the challenges of that culture. It is part of our culture to recognise at last our cognitive precariousness (*ibid.* p. xvii).

4. TRUTH

We have seen how this challenge has been taken up in the social history of science, and how it resituates scientific theory in the mutually interacting and holistic complex of social myth. What, however, of the consequent threat of cultural relativism? Durkheim sought to evade the threat not only by anchoring knowledge in the absolute objectivity of natural science, but also in a faith in evolution.[7] Recognising, in spite of his methodological reliance on science, that basic morality is not subject to scientific decision, he developed a theory of ethics in which, in the long run and with possible historical vicissitudes, what evolves in the history of societies is by definition good. Such a theory is hardly compatible as it stands with belief in the good as the will of God, but we may take a hint from it in suggesting the correct religious response to relativism. If the various arguments from philosophy and history of science, and from the sociology of religion, are rightly taken as overwhelming support for a thesis of the cultural relativity of knowledge claims, then there is no *natural logos* or *telos* to which we can appeal beyond relativity. But religious belief is belief in a supernatural *logos* and an extra-spatiotemporal *telos*. We must look to belief in God for the providence that underlies our shifting cultural schemata, and for the inspiration for countless cognitive decisions that are taken within them in history.

Returning from such ultimate considerations, however, there is more to be said about the nature of truth. I said earlier that the forms of truth required for understanding both scientific theory and religious doctrine must be other than the 'objective' truth of instrumental science which is derived from its success in empirical prediction and control. It is this truth of 'objective' fact that is exclusively accounted for by the currently received philosophical theory of literal or univocal meanings and correspondence with natural reality. In our account of the interaction of cognitive systems of different kinds, however, we need a quite different theory of truth which will be characterised by *consensus* and *coherence* rather than correspondence, by *holism* of meanings rather than atomism, by *metaphor* and *symbol* rather than literalism and univocity, by intrinsic judgments of *value* as well as of fact. Development of such a new theory of language, meaning and truth is a long-term project for philosophy,[8] and until it is attempted we shall have no satisfactory understanding of the epistemological significance of either scientific theory or religious doctrine or of the potential conflicts between them.

Meanwhile, however, these considerations do shed light on the present status of the conflict between cosmology and Christian doctrines of creation. As at present conducted, they are like different language-games, having their own methodologies and their own criteria of meaning, truth and value. But we have seen that conflicting mythologies should not be seen as autonomous atoms within the social milieu, but as interacting systems of cognition and value. This means that judgments that are most obvious on 'scientific' grounds are not always those that are most appropriate in given historical circumstances. For example, posterity has found it easy to champion Galileo against the Church. But this was a dispute between, on the one hand, a cosmological model that yielded more or less correct planetary motions, against, on the other hand, an Aristotelian and Ptolemaic myth which underpinned theological interpretations of man in the universe. Such a myth may under particular circumstances have as good or better claims to allegiance as a less comprehensive cosmology. And note that the Copernican model was not 'objectively true' either.

Notes

1. For further discussion of this point, see W. O. Quine *Ontological Relativity* (New York 1969), and my *Revolutions and Reconstructions in the Philosophy of Science* (Brighton, Sussex 1980) chapters 6 and 8. For arguments against scientific realism, see B. Van Fraassen *The Scientific Image* (Oxford 1980).

2. For a more positive view of art and play, see H.-G. Gadamer *Truth and Method* (London 1975) (*Wahrheit und Methode,* Tübingen, 1960), First Part.

3. 'Physics and psychics: Science, symbolic action, and social control in late Victorian England' *Natural Order*, ed. B. Barnes and S. Shapin (London 1979) p. 167. For further discussion and references to work in the social history of science, see my *Revolutions and Reconstructions*, chapters 1 and 2.

4. 'Weimar culture, causality and quantum theory 1918-1927: adaptation by German physicists and mathematicians to a hostile environment' *Historical Studies in Physical Science*, ed. R. McCormmach, 3, 1971, 1.

5. London, 1915 (*Les Formes élémentaires de la vie religieuse,* Paris, 1912).

6. London, 1963 (*De Quelques formes primitives de classification, Année Sociologique,* 1901-1902, Paris, 1903).

7. *Sociology and Philosophy* (New York 1974) (*Sociologie et Philosophie,* Paris, 1924).

8. For useful stating-points of this project, see, for example, G. Lakoff and M. Johnson *Metaphors We Live By* (Chicago 1980) and J. Ross *Portraying Analogy* (Cambridge 1982).

Langdon Gilkey

The Creationist Issue: A Theologian's View

INTRODUCTION

THE RECENT court trial in Arkansas concerning Creation-Science and Evolution—and the upcoming trial this fall in Louisiana—have raised surprising and important issues for both the scientific and the religious communities of America. They force us again to think about what the traditional separation of Church and State means in modern life, and anew about the vast complexity of the relation of science to religion in advanced technological culture. Most people, both in the scientific community and among the media, picture this controversy as simply the latest act in the age-old and continuing drama entitled, 'The Warfare of Science and Religion'. That this is a serious misreading of the controversy, I shall here try to show: there is (as there was in the nineteenth century) a good deal of 'science' and a good deal of 'religion' on *both* sides of the case. The controversy can, therefore, best be seen as representing a contest between *two different sorts* of interrelations, or of union, of science and religion.

As this article will try to suggest, this controversy has a wider application than merely to the North American continent. To be sure, that continent represents a different religious *'Gestalt'* than do Britain and Europe. As is well known, concern for and participation in religion are very much more widespread, and, as a consequence, aggressively conservative and fundamentalist forms of Christianity represent much more of an active, shaping and perhaps threatening social reality than across the Atlantic. For this reason European observers are for many good reasons apt to wonder not only how such a bizarre and anachronistic conflict could occur in the so-called 'leader' of the scientific twentieth century, but to be also quietly grateful that such 'throw backs' to the pre-scientific age do not occur in older, wiser and so more moderate Europe. A couple of remarks might be made in this connection, not so much to chide this reaction as to set the event itself in perspective.

First of all, the North American continent represents a consciously and deliberately, even belligerently, *non-hierarchical* society. There are vastly different levels of wealth, of power and of social 'elegance', of course, and differences of intellectual, cultural and academic level are as pervasive and evident as elsewhere. We, too, have our élites, and they are not only as brilliant and beautiful but also as smug, snobbish and arrogant as elsewhere. The point is they have little if any cultural or spiritual authority in the rest of the culture as a whole; in fact other intellectual and spiritual levels would *resist* their authority with passion. This is perhaps especially true in religious circles. North American culture is dominated by a free-church, congregational consciousness and by a wider religious pluralism unfelt in Europe. No one group or communion speaks for any other one or exercises much influence over any other. Thus while there are spiritual hierarchies with some authority (though not much!) within their own religious community—as among the Roman Catholics, the Episcopalians and the Lutherans— such hierarchies have virtually no influence at all outside the bonds of that communion. And most of American and Canadian Christianity exists outside the bounds of such hierarchies in individual free churches and in a vast number of different denominations. Thus whereas in Europe most of church life takes place *within* and so *under* the authority of a given hierarchy, Catholic or Protestant; and established hierarchies are always ruled by the cultural if not the economic or political élite, in contrast, little of American church life exists within such an established hierarchy; and for none of those outside do the hierarchical leaders (ecclesiastical or theological) or the main-lines churches have an authority at all.

In sum, while in Europe the ecclesiastical and theological élites speak generally for all the churches and churchpeople, and mould public interpretations of Christianity up and down the various levels of society, by contrast in North America they do not. In fact such élites would be regarded here not only as effete and worldly but also as hopeless compromisers with the world's various cultural and social errors. There is unquestionably much non-élite, anachronistic and 'backward' religion in Europe, in Cornwall and Yorkshire, in rural France, in Bavaria, in Spain and in Sicily; but in no case is one of these sorts of 'popular piety' *on its own*. Rather those who represent and speak for it in the wider public life are the élite bishops and archbishops in London, Paris, Munich, Madrid and Rome. Popular piety abounds there, too, but on a local, well-controlled level. Here popular religion is on its own; it deliberately distances itself from these élites; and thus popular piety, already more active and pervasive, takes on a social and political reality unknown in more traditional and so hierarchical lands where most forms of cultural and spiritual power remain in the hands of the culturally élite.

Despite these large and important sociological differences in the role of religion and in the distribution of power within religious groups, the most fundamental anatomy of this controversy is by no means strange to Europe. As I shall try to show, each side in this controversy represents *one* sort of union of science and the religious: on the one hand, a 'popular' sort where popular religion and the newly-born *popular* science unite into the bizarre hybrid of creation science; and on the other hand, the more respectable union of élite science and élite religion represented by the AAAS and the National Council of Churches (or the Catholic Theological Society!)—and at Little Rock by the 'religious' and the 'scientific' witnesses against the creationist law. As we have noted, the creationist union of fundamentalist religion and popular science is in a sense typically American. But Europe also has had its unities of science and the religious, fully as bizarre and menacing as this one. Fascism and Nazism were examples: there science and technology made union with and became instruments of ideologies with religious overtones—and there science was as strangely misunderstood and misused as in the present case. Some might feel the same of the science and technology present in the Universities of Moscow, Prague and East Berlin, each comparable in many important respects to our Moral Majority.

These examples of 'religious ideology' are not traditional, nor are they forms of traditional religion. But the anatomy is the same: a union of modern science, technology and an ultimate, sacred and all-determining symbolic universe. And each represents a very different sort of unity from the liberal amalgam of Enlightenment science and enlightened religion of which we élites may approve. Our controversy, therefore, may be more relevant to the European scene, past and future, than seems at first glance to be the case.

1. CREATION-SCIENCE: AN UNEXPECTED HYBRID OF SCIENCE AND RELIGION

Let us begin by stating the definition of Creation-Science and of Evolution-Science as the Creationist movement views them, and as they were expressed in Act 590 in Arkansas. Creation-Science represents the following points; the similarity, if not the identity, of these points to those in the Genesis account taken literally, is, of course, immediately evident. (1) Sudden creation of all things (universe, life, man) from nothing. (2) Permanent species or 'kinds of things', going back to the very beginning. (3) Separate ancestry of apes and of man. (4) The explanation of geological formations and changes by means of 'catastrophes', for example, the flood of Noah. (5) Recent creation of all things (namely, within the last 10,000 to 25,000 years). As Creation-Science documents admitted, such a model requires a Creator to be intelligible. The model is, therefore, 'scientific' as representing a rational explanation of what was regarded as 'scientific facts', namely, as we shall see, facts associated with the above five points. Contrasted with this model for the Creationists is the other model, named Evolution-Science. It specifies that the origin of the universe, of earth, of life and of man lay in 'natural forces' alone, in blind matter ruled by impersonal laws. Since, the argument continues, origins cannot, in either one of these cases, be 'observed', evolution is hardly scientific, strictly speaking. Thus evolution represents an atheistic and humanistic 'religion' (the belief that everything arises out of blind matter) fully as much as Creation-Science represents biblical religion, the belief in Creation by God. As a consequence, these two opposing models or theories are, at best, 'equally scientific', and 'equally religious'. Hence, if one is taught, it is only fair that the other also be taught, and it is precisely this balanced treatment that the laws require.

The laws seeking to establish the teaching of Creation-Science so defined alongside of Evolution-Science in the public schools have understandably enjoyed a good deal of wide public approval and support. On the surface, they seem—even to those quite uninterested in, or even opposed to, Creationism as a belief—eminently fair. They state the proposition—by no means incredible to the public at large—that these are both 'scientific models of origins'. (Do not large numbers of Americans regard each of these as true; and, if they are true, are they not thereby scientific?) And they claim further that, since these are *the only two* explanations of origins (and who, in the general public, knows this to be false?), should they not, as the two available alternatives, be given equal time or balanced treatment?

Our remarks in this article, therefore, will seek to show that these two claims are, in fact, false: (1) Creation-Science is *not* a scientific model, and therefore it is not at all a direct alternative to the scientific theory of evolution. On the contrary, it represents a religious or theological model of the explanation of origins, which neither conflicts with nor excludes scientific theories of origins. To state in legislation, as these laws do, that these are alternative models, categorically opposed to one another, is for the State officially to sow a double confusion and to promulgate a serious error: first, confusion about what science is and about what religious explanation is, and second, the error that religious and scientific views of origins are unalterably opposed to one another, so that no one can be a Christian believer and still accept evolution. Here the State would explicitly set itself *against* the forms of belief characteristic of most of the nation's churches and synagogues, not to mention those with other religious orientations.

(2) Creation-Science and evolution do *not* represent the 'only two explanations of origins', contrary to what every Creation-Science volume and each of the two laws states. There are innumerable theories of origins, of which these are only two. Each

religious tradition in history has had a different 'myth', doctrine or truth about how things came to be; and today the understanding of origins differs markedly between different religious groups, between Hindus, Buddhists, Christians, Jews and Moslems, to name only the largest groups. Among the many views of origins in the history of religions, Creation-Science, therefore, represents one rather minor interpretation or variant of the Christian view, an interpretation not even shared by most present-day Christian groups or churches, and certainly not shared by the Jews, Moslems, Hindus, Buddhists, Sikhs, and American Indians who are also American citizens. For the State to require the teaching of Creation-Science is, therefore, to legislate not only that a *religious* doctrine be taught as science, but even more, that a *particular* religious doctrine, which excludes vast numbers of other religious believers, as well as non-believers, be taught to the exclusion of other religious views.

The stated purpose of these laws, namely to give 'each of the two alternative scientific models' balanced treatment, turns out, therefore, on examination to represent a very misleading and, in fact, unfair enterprise indeed. These are *not* alternative scientific explanations of origins at all, but theories on quite different levels, the one scientific and the other religious. Nor are they the only two theories either extant or relevant. For the State to require the teaching of only one of the many current religious theories implies officially that, if anyone be religious or Christian, they must agree with Creation-Science and be against evolution; or, alternatively, if they support rather than oppose evolution, they must be atheistic, as the Creationists say evolutionists inevitably are. Non-fundamentalist religion, Christian or non-Christian, as well as science, is put in jeopardy by these laws—which is why there were in Arkansas more religious groups opposing the law than scientific groups or leaders.

To support the above claims against Creation-Science (that it is religion and not science, and that it represents only one among innumerable religious views), let us examine their description of this model more closely. The moment one reads the documents (books, pamphlets, journals, even comic books) or listens to the expositors of Creation-Science, one encounters an unexpected, even paradoxical, conjunction of apparent opposites: these are 'scientists' (most have reputable doctorates in theoretical science) who are also fundamentalist Christians. Accordingly, their theory represents a literalistic version of Genesis presented as 'science' or as 'a scientific model', a 'sport' in a developing scientific culture if ever there was one!

Earlier anti-evolutionary movements, in contrast, were explicitly anti-scientific; and they argued in public forums and in the courts against science and for 'Biblical Christianity'. By the last decades of the twentieth century, however, much has changed; for the culture has become scientific and technological from top to bottom. Now we find anti-evolutionary groups claiming to represent 'science' and not religion; consequently, they attack evolution not as science, but as 'bad science', and they defend their views of origins as 'good science' and thus as quite independent of any religious sources. Just as fundamentalist groups have recently adopted, and now largely dominate, much local television technology, innumerable commercial enterprises, and a number of brand-new 'universities', so fundamentalist intellectuals have recently entered the realm of science, become thoroughly trained there, absorbed much of its content and know-how, and now seek to shape its inquiry and teaching in their own doctrinal directions. As we noted, they represent not so much the antagonism of religion against science as a new sort of *union* of religion (fundamentalist religion) both with technological expertise and with theoretical science at the doctoral level.

Let us see, then, how it is that these 'scientists' can present what is clearly a literalistic account of the Creation story in Genesis as a legitimate 'scientific model'—for surely that would represent a neat trick, if it could be done. First of all, they define science in terms of 'scientific facts' or 'scientific evidences'; it is, say they, the scientific facts that it explains that makes any theory science. This makes sense to most of us who think of scientific discovery as the discovery of facts, of science as an accumulation of facts, and so a scientific theory as the reasonable explanation of such facts. (It also makes sense to a culture that sets examinations in science, in medicine, and even in the liberal arts that are exclusively true-false or factual discriminations!) As these many examples show, a scientific culture becomes rapidly a culture that worships facts as embodying all truth, and that gradually loses its awareness of and interest in the importance of *theory*; it sinks from a scientific culture to a technological one. The fundamentalist scientists are no exceptions to this general cultural characteristic of our time.

The 'facts' to which they generally point in their arguments are threefold: (1) Evidences of design, order, and purpose in nature that argue for an intelligent and purposive Creator; such evidences of order have been, of course, uncovered in abundance by all manner of scientific inquiry. (2) Arguments against prevailing geological and biological theories or hypotheses (and astronomical ones, when the age or the expansion of the universe is at issue) developed by other scientists (who accept evolution as a 'fact' but argue with their colleagues' theories explaining it). Such arguments against geological and evolutionary theories do not, of course, count at all as arguments *for* Creationism—unless one accepts (as these authors and their readers do) that these are the only two possible past, present or future theories of origins; otherwise, to disprove theory A is certainly not to establish theory B. (3) Various attempts are made, following these latter arguments, to show that the Creation hypothesis 'predicts' certain (carefully chosen) 'facts of science' better than do evolutionary theories. To this layman in science, these predictions have little to do with experimental prediction in science—but that represents a point too complex to outline here.

In any case, the only 'good' arguments in this list—and also the oldest by far—are the first, arguments for a Creator-God from the evident design of the world. From Greek and Roman times to our own, these have seemed to many to be very convincing arguments, and as a result, they have constituted most of the traditional arguments of 'natural theology', that is, philosophical arguments for God's existence based on no religious authority, religious sources or religious experiences, but on reason and experience alone. The only trouble here is that these are not—and never have been considered to be—*scientific* arguments. To some commentators, they represent at best speculative philosophical arguments; to other, less enthusiastic interpreters, they are only pale rationalisations of religious faith—to none, until the present (when science became the main model of truth), were they ever thought to be scientific. Whether or not the Creation-Scientists who have advanced them as science were aware of these points or cared about them, in any case they have represented them as scientific arguments, and as 'scientific evidences for Creation and scientific inferences therefrom'—and so it is on this ground that the traditional (literalistic) doctrine of Creation is now presented as a scientific model.

Clearly, there is a confusion here; but where, precisely, is it that they are wrong? Why are these not scientific arguments; why is not their implication or conclusion: that God created the entire realm of nature (true as it well may be), a *scientific* conclusion, a legitimate part of recognised scientific theory? The central error involved here stems from their initial (and surprising, since they are trained in science) misunderstanding of science. Science is not located in its facts; thus scientific theories do not represent merely 'sensible' or 'rational' explanations of scientific facts. On the contrary, science is located in its *theories*; it is the *theoretical structure*, the coherent system of theories, created by scientific inquiry, that constitutes science—not the facts associated with those theories. In other words, it is the *way* science explains facts, not the *kinds of facts* it explains, that makes science science. Further, these theories that constitute scientific explanations are of a certain sort; not just any old theory is scientific, even if it gives a 'reasonable' explanation of data of which we are certain. There have been all sorts of 'explanations': magical, superstitious, traditional, religious, philosophical, common-sense, of the recognised 'facts' of experience however reasonable they may seem, none of these are thereby science. A theory must conform to the methods and canons of science to be science—an aspect of science well known since at least the seventeenth century but obscured in a technological culture, where science is established and therefore not much thought about (as much important knowledge about Christianity was thoroughly obscured by the establishment of Christianity in 'Christendom'. Establishment has always endangered the clarity and purity of a religion; it seems it may similarly endanger science).

Basic to these 'canons' or 'rules of the road' of scientific procedure are three requirements; if it defies any of these, a theory cannot be said to be scientific. (1) No supernatural agent or force can be appealed to in the explanation; on the contrary, explanations must be in terms of forces or factors that are *finite*, and in that sense 'natural' (forces of nature, trends in history, human actions, etc.). (2) Explanations must be in terms of natural laws, that is, *invariable* and *necessary* forces or factors. A scientific explanation must be necessary and universal (if P, *then* Q) if it is to be fully scientific; hence natural science does not explain in terms of *unique* actions, *purposive* actions or intentions (e.g., God created out of love). (3) Scientific theories must *grow out of* experimental evidence (what they refer to must be locatable within it) and *be tested* by further experiment (shareable and public) evidence. Clearly, a theory or model that refers to the action of a transcendent Creator establishing in a unique act at the beginning of time the entire realm of nature, and so the system itself of natural laws, defies each of these canons. It appeals to a transcendent, divine power far beyond (because the source of) the system of finite causes; it explains in terms of the power and purpose of that Creator, not in terms of natural laws (these are here established and so 'not yet' in effect); and being a unique, unrepeatable act, it cannot be tested in the present, since there are no similar processes available in present experience. Whatever facts it may point to, the model of Creation cannot represent a scientific theory, since it directly contravenes all the requirements that make any theory scientific. Of course (at least to religious—and many other—minds), the fact 'creation by a divine being' may not represent a scientific hypothesis or theory does not mean it is not *true*, unless scientific theories are regarded as exhausting all relevant truth. It does mean, however, that it represents some other mode of cognition or of truth.

The ACLU case in the trial, therefore, hung on the point that the affirmation of the creation of the universe by God, common to the Christian and Jewish religions, represents at base a 'religious' type of truth, and consequently, the model of Creation-Science (as a literalistic variant of that tradition) represents a religious type of explanation. To prove that it was not science (as we have just briefly done here) was, to be sure, both interesting and important; but it is not unconstitutional to mandate a theory that turns out not to be scientific, or even not to be true. It *is* unconstitutional, however, to mandate in law the teaching of a particular religious doctrine or theory. Thus, to show that this model was in fact a religious theory represented the centre of the plaintiff's case.[1]

As we have noted, the dependence of Creation-Science on the Genesis account is crystal-clear, as is the adherence of the authors and proponents of Creation-Science to fundamentalist Chistian religious groups. Still, these relations had to be spelled out precisely and in some detail to make our case in court. Thus part of our testimony (that of Fr Bruce Vawter) established the connection of Creation-Science as defined in the law to the Hebrew text of Genesis. Another witness (Professor George Marsden of Calvin College) showed the source and development of these ideas in the history of American Protestant fundamentalism in the twentieth century. And finally, Professor

Dorothy Nelkin of Cornell established the relation of Act 590 to current fundamentalist groups and movements. My own testimony, as a theologian, took another tack: my aim was to show that the model or theory enunciated by Creation-Science was essentially and innately a religious theory. Whatever it said it was, and whether or not it appealed to Scripture, church authority or church doctrine, it could not, by the nature of its very conceptuality, avoid being an example of religious speech, an expression of religious belief.

Religion—what is religion and what is not—must, I said, be defined historically. This means that while many religions (and so many cultures) do not speak of God, still in the West, dominated as it has been by the Jewish and the Christian traditions, 'God' is the centre of all religion. This is the functional meaning of monotheism (or the First Commandment): all that is religious concerns God, and God alone. Any worship, anything 'religious' that is not related to God, is, therefore, blasphemy or heresy, out of court in such a religious culture. Just as all that is religious is related to God, so as a consequence, all that is related to God is religious. God's being and God's acts form the centre of the witness, the concern, and the worship of these religious traditions, and the recital of these 'acts' represents the main element of their religious doctrines or religious truths. This recital includes *all* of God's actions, from Creation, through Revelation, the giving of the Law, the Incarnation, the final Redemption—as the Scriptures and the Creeds that begin with the first and end with the last illustrate. It is no accident, then, that the affirmation of divine Creation is found in the first book of the Jewish and the Christian Scriptures, and represents the first article of the oldest Christian Creed ('I believe in God the Father Almighty, Maker of Heaven and Earth'). In Judaism and in Christianity, therefore, the belief in God the Creator is the first and basic religious belief characteristic of these traditions. To speak of God—the transcendent God who creates and redeems—is to speak religiously.[2]

If, moreover, to speak of *God* in this context is to speak religiously, then the theory of creation out of nothing (traditional to Christianity and asserted by the Creationists) represents the quintessence of religious speech or of a religious concept. For here all other forces and factors are absent. Since every other agent appears *because of* and so *after* this act, in referring to this act we refer *only* to God: 'God created all else out of nothing'. As I remarked at the trial: 'This act is even more religious than Christmas (gasps of surprise), since there, however one interprets it, Mary was *also* present.' This means not only that Creation represents the epitome of religious speech. Even more, it means that without God, the transcendent Creator, there can be no creative act—for no other creative force is present. Without God as agent, therefore, there is no theory or model here at all. Yet with God, the model becomes essentially and inevitably a religious theory. Thus if it *is* a theory, and is internally coherent and meaningful, it is a religious theory; if it abstracts God out, and thus seeks to avoid being religious, it represents no concept at all, for it is void and meaningless. The Creation-Science model is either religious or it is nothing at all.

Not only is this concept of creation (God brings all into being out of nothing) quintessential to religious speech; it also represents, we argued, a *particular* religious viewpoint on origins. As we noted, there are many models of origins in religious history: in some, a god 'creates' (or makes) the world out of pre-existent matter (or from the body of a slain monster); in some, several gods co-operate; in some, the world arises *out of* the god by emanation (or birth) and returns to the divine at the end of the age; in some, it is always there, and so had no beginning at all. None of these can, in the wildest stretch of the imagination, be called 'evolutionary' theories; nor are any of them similar to the Christian view of a self-sufficient transcendent deity who, through his power and will, brings all into existence 'out of nothing'. This latter concept is, therefore, unique to the Christian (and possibly the Jewish and the Islamic) tradition; its literal interpretation is confined to fundamentalist wings of that one tradition. While it is certainly religious, Creation-Science does not even represent 'religion in general' or a universal religious position—as is also claimed.[3] It represents a particular minor viewpoint in one religious tradition. Its establishment would be a manifest wrong to all the other religious traditions in our land.

Returning to our analysis of Creation-Science, it is evident that it opposed 'Evolution-Science' on two interrelated but distinguishable grounds. The first could be called religious; it saw in evolutionary theory a religious perspective—which it called atheistic humanism—antithetical to Christian faith and its reverence for the Bible. Thus it termed evolutionary-science 'a religion', a materialistic, godless, amoral faith, the source, in fact, of most of the present evils of our history: Communism, nihilism, relativism, liberalism, ending with homosexuality, ERA, the UN, and the National Council of Churches! To believe in evolution, so they affirm, is to adopt, perhaps unwittingly, this 'religion'; it is, at the least, to be a fellow-traveller, an involuntary agent of religious atheism. That there is in more ways than one a *connection* (if not an identity) between modern science and modern evolutionary-science on the one hand, and 'natural humanism' on the other, there can be little doubt—here the Creationists are right. But that there is also a *distinction* between the scientific theory, or theories, of evolution and such a naturalistic, atheistic, humanistic philosophy or religion is also true. Part of our purpose here is to make both that connection and that distinction clearer than they have made it.

In any case, to them—and here, they were right—part of the struggle represented in this trial was a contest between what they termed two 'religions', two religious perspectives on all of existence, and so on the character, responsibilities and destiny of human existence. One of them is biblical and Christian and centres itself on God the Creator and Redeemer; the other is naturalistic, atheistic and humanistic, and centres its hopes on human being, its powers, its intelligence, and its goodness. The scientific community, or much of it, denies this 'religious aura' associated with much of modern science; it sees itself as purely objective and scientific and freed from all taint of religion (see the writings of Julian Huxley, George Gaylord Simpson, Bronowski, and Carl Sagan). In fact, however, they and their followers assume an *identity* between 'evolutionary science', as they understand it, and the naturalistic humanism they also profess—and they argue this identity in every volume they write or address they deliver. The Creationists have seen this, and they achieve much of their credibility by pointing it out.

But there is more to Creationism than this, and their antipathy to evolution has other bases than this 'religious' one. Just as many in the scientific community have identified their understanding of science with a naturalistic religious perspective, so the Creationists identify their religious symbols or doctrines with certain 'truths' in the area of earthly and empirical matters of fact, that is, in the areas with which science as science (and *not* as 'religion') is concerned. Thus their religious doctrines include authoritative ('revealed') statements about the age of the earth, whether the stars have changed or not, how the earth developed, how the different kinds of living things arose, and so forth, issues with which the natural and historical sciences directly deal. As many scientists see science to be capable of generating a true religious or philosophical perspective, so the Creationists believe revelation provides us with a divine astronomy, geology, biology, and botany—not to mention the physics and the chemistry that underlie these other sciences. These matters represent, however, questions concerning relations among *finite* things, an understanding of the structure, the history, the causes, and the outcomes of natural and historical events. This is *not* the area with which religion provides authoritative information; religion concerns the activity of God in and through events, but it does not inform us in detail about the events themselves, taken merely on their own, and as we can know them empirically. This is the area within which science (and history) have developed competence, and which, we now know, they are competent provisionally, if not finally, to understand and to interpret. That the Creationists *also* assent to and accept this 'competence', their participation in science and in its technological results clearly manifests. Here they are utterly inconsistent: the scientific knowledge of which their writings are full, and the technology they so cheerfully use, both depend on that competence, on the scientific method that enacts that competence, and so on the very conclusions of scientific inquiry they here deny.

The deepest problem of Creationism, therefore—the hard 'nut' that seems quite uncrackable—is that they are, on religious grounds, opposed not only to modern evolutionary science as 'religion', but to the most pervasive, important, and general substantive theorem of contemporary science (of modern science as *science*), namely, to the proposition that the natural universe we inhabit is eons old, has undergone innumerable basic changes in its process of development (e.g., the stars have changed in many significant ways, a point they deny), and that new forms of stellar and galactic entities, of the earth itself, and new species of marine, of plant, of animal, and of human life have developed in this long process. To deny this is to deny the central *content* of most of our sciences, and to reject the theorem on which our most important inquiries are based, the theoretical foundation of our present technology and industry, not to say our academic culture. Here is perhaps the fundamental point at issue: though they are quite right about the predominantly secularistic and even atheistic character of most of our contemporary intellectual life, they are (so most religious groups feel), quite wrong to deny and refute, in the name of religion, the entire scientific understanding of nature and of its history on which rests much of the civilisation they also enjoy.

If one seeks for the deeper source of the error involved in this confusion of religion with science, of the religious truths about God and His Creation, which they treasure, with the scientific 'truths' about the history of nature which they reject, it lies, I believe, in the assumption that there is only *one* kind of truth, that truth exists, so to speak, all on one level. This one relevant sort of truth is material, physical, and historical, 'factual truth', even a quantitative truth: the age of things, when they came to be, how they came to be, what causes brought them about, and so on. This sort of information, for them, science tells us; this sort of information God—or revelation—also tells us. If God created us, He must have acted like any other physical cause, and thus in acting, replaced all other physical causes. Because these truths (scientific and religious) are of the same sort, therefore when each of them tells a different story, then of course they contradict and so exclude one another. Thus does a religious theory of origins conflict with a scientific theory of origins—for they both deal with reality (or the origins of reality) *in the same way and on the same level*. It is this view of truth as all of one kind, as material, historical, and factual truth—truth about a physical event one might observe or a photographer capture—that leads to the deepest error of Creation-Science and its continuing antipathy to liberal religion and to scientific education.

It is also clear that this same error is not confined to the fundamentalist community. It is shared by many in the academic and the scientific communities—by all those who view religion as merely 'pre-scientific' theories about the earth and its origins, and so who view scientific knowledge as replacing so-called religious knowledge; all those who say, 'we scientists now know there is no God'. It is because of this error that many scientists, sure that traditional religious faith in God has been dissolved by scientific inquiry, assume that with the developments of science, a naturalistic philosophical or religious perspective must replace a traditional Christian or Jewish one. Ironically, the same error, typical of a scientific and a technological culture (can one imagine a medieval Christian or a medieval Buddhist, educated or uneducated, thinking *factual physical* truth was everything?), pervades both sides of this controversy, the

Creationists on the one hand and many of their scientific opponents on the other. Truly this controversy has been bred by the character of *both* the science and the religion in a scientific and technological culture.

2. RELIGION AND SCIENCE IN A TECHNOLOGICAL CULTURE

As both sides of the Creationist controversy have illustrated, the relations of science to the religious in a technological age are complex indeed. Since there is a surprising amount of science, as well as fundamentalist religion, on the side of Creation-Science, and more than a little of the 'religious' on the side of the established scientific community, this is no simple warfare between advancing forces of scientific light and retreating forces of religious darkness. It is now time, therefore, to look more closely at the complicated relations of science and the religious in an advanced culture. Are there possibilities of a *more creative sort of union between science and the religious* than is offered to us in Creation-Science?

(1) The first area of this complex situation that needs clarification is what may be termed the 'methodological non-theism' of scientific inquiry. As we have noted, scientific explanations refer to or use only natural, finite, or 'earthly' causes to explain the processes and changes of nature. When science explains a process of changes, or of growth and development (or of decay and disintegration), it explains it by preceding causes which are themselves part of the same stream of experience, causes, that is, on the same level; it asks how did A arise out of B and C, B and C perforce being in principle as experienceable by the senses as is A. Two basic presuppositions are latent in this fundamental rule of scientific method. (i) As a condition of its inquiry, science presupposes a preceding 'situation' and its causes out of which the events or processes of change it investigates arose. For example, it presupposes the existence of mothers and fathers when it inquires into the birth of a child (as Aristotle stated this basic rule: 'Nothing can come from nothing'). Science cannot, therefore, inquire into, or know about, an 'absolute' beginning, an origin of the *entire* system of finite, natural causes.

(ii) Science is *confined to* the system of finite causes; it cannot search for or recognise—and so refer to—a trans-natural or supernatural cause that has intervened into the sequence of natural events (a miracle). As David Hume rightly argued,[4] the commitment of the scientist is to find the natural cause. Even, for example, if he or she believes that the 'hand of God' was at work in a cure (or a disease), nevertheless, the medical researcher is obligated as a scientist to search for and locate the preceding *natural* cause of either one; for only then are medical knowledge and medical ability enlarged. Miracles (as well as absolute beginnings) may well be there, but scientific inquiry, because of its methodological commitments, cannot know of them. It is confined to the sequence of natural or finite causes, to the factors within the stream of ordinary experience.

As a consequence of these two presuppositions, science can, and does, inquire into how something arose out of *something else*, what can be called the question of *proximate origins*; it cannot seek or question beyond the powers and factors discoverable to sensory, and so to common, investigation. It cannot, and does not, inquire into *ultimate origins*: the question, where did the entire system of finite causes come from, what is its ultimate source or ground? Or, as the Scholastics put this, science deals with *secondary* causes alone (creaturely causes); it cannot deal with the *primary* causality of God which brought all secondary causes into existence, maintains them there and works through them. If, then, God be that ultimate source and ground, the primary cause transcendent to the system of finite causes because God is their origin, then God, so conceived, *cannot* be part of a scientific hypothesis. Science is *methodologically* non-theological or non-theistic; it *cannot* raise the question of God, no matter how 'religious' the scientist in question is, or how firmly he or she believes the presence of the divine power and wisdom to be there.

This point is very important in this controversy. The Creationists have stated that since evolutionary science explains 'origins' without reference to God, *therefore* it is naturalistic, atheistic, and humanistic. Clearly this is an error, a serious misunderstanding of the character and so the limits of scientific inquiry. To say that a scientific theory of origins makes no mention of God is merely to say that it is scientific; there is nothing at all intrinsically atheistic about a scientific hypothesis—unless the scientist wishes to go *beyond* science to state a naturalistic philosophical and/or religious view of the universe as a whole. But, like its step-sister, Creation-Science, that naturalistic 'extension beyond science' is not scientific.

Lest natural science seem by this account to be suspiciously secular in character, let us recall that in modern culture historical inquiry (history as a study) and the law alike reflect the same restriction to finite, 'earthly' causes. If I write a *historical* account of the origins of the Second World War, and hope to have it received as a historical interpretation by the community of historians, I must confine my discussion to such causes as economic factors, historical trends, political relations, the intentions and acts of statesmen, and so on. I cannot appeal to 'the judgment of God on the empires of England and of Europe' as an explanation—any more than I can explain Constantine's defeat of his brother in terms of the host of angels at the famous bridge. However, as a theologian giving a *theological* interpretation of history, I can speak (and have spoken[5]) intelligibly and persuasively about the work of God in history. Similarly, as a lawyer defending a client from the charge of murder, I cannot in a modern court (even in Arkansas) advance a theory of the crime that makes God, and not my client, responsible for the death. And this does not mean that lawyer, client, or judge and jury are atheistic, or that the State of Arkansas is atheistic. It simply means that our present law, like our historical inquiry and our natural science, proceeds as a 'secular' discipline, confined by its own methodological presuppositions to natural and historical (secondary or finite) causes, and it can in its procedures only use such causes in the construction of its theories.

This inherent limitation on scientific inquiry—that it can only deal with preceding finite (natural and historical) secondary causes discoverable by empirical inquiry (discoverable to sense)—separates or distinguishes scientific inquiry in itself from any general view of the whole of things, any total or global interpretation of reality as a whole—and so from any religious questions of 'ultimate origins' (or of primary causality). Scientific theories are thus significantly different from metaphysical and/or theological views of all of reality, of reality as a whole. When, therefore, science offers *its* explanation of origins, it presents a limited explanation of reality viewed from a distinct and restricted perspective. Its theories, therefore, propose no view at all about the ultimate origin, ground, or source of the cosmos, of life, or of humankind—any more than a biological, zoological, physiological or medical explanation of *my* origin in my parents and grandparents excludes a metaphysical or religious explanation of my origins and destiny in God. As noted, evolutionary theories, as scientific, are consequently not in themselves atheistic (as both Creationists and many scientists together maintain). By the same token, astronomical, geological and biological theories of origins do not at all exclude a *theistic* explanation or theory of the ultimate origin and ground of the physical processes examined by these special sciences, processes through which God is believed to have worked. It is the task of theology (especially of philosophical theology) to explore how a theistic interpretation of reality as a whole—and so of human, as well as nature's, origins and destiny—can be developed which accords with and supports, rather than contradicts, the scientific view of the natural world and the historical view of our common history which the 'scientific' culture that the theologian inhabits assumes in some sense to be valid.

Such theological syntheses of modern science with the Christian and the Jewish belief of God are not only theoretically possible—as the above analysis shows—they have also been actual. Any number of them have appeared since modern natural science began in the seventeenth and eighteenth centuries, and even more since the advent of evolutionary theories in the nineteenth century. While some of these, to be sure, are more profound, coherent, and persuasive than others, still it is evident that a theistic interpretation of evolution—or (better) an evolutionary interpretation of theism—is perfectly coherent and, so I believe, a good deal more rationally, as well as religiously, persuasive than is its naturalistic philosophical counterpart. These theological syntheses, as well as a critical and symbolic interpretation of the Book of Genesis, have been central staples in the education of theologians, ministers, and rabbis (and now, latterly, of priests) for over 100 years. That so many persons in our present culture, those inside as well as outside the major churches and synagogues, should be unaware of this accord between modern scientific theories of development and faith in God as Creator is thus a large responsibility of the churches, and especially of their leadership. Known well by every clergyman and accepted by most, this accord with science is still perhaps the best-kept secret in current American religion. Thus have the churches unwittingly helped the Creationist cause; for the majority of the public accepts as valid the Creationist claim that 'if you believe in evolution, you must have already abandoned a Christian (or Jewish) belief in Creation'.

(2) Although scientific inquiry in itself leads by no empirical or logical necessity to any particular philosophical or religious world view, still it is clearly possible for science—as an attitude, a method, and a body of conclusions—to 'expand' itself into a view of the whole of reality, into a naturalistic philosophy and so a humanistic religion. A scientific culture produces its own 'myths' just as a fishing culture does; and it is quite natural that as the myths of a fishing culture have a base in fishing, so in a scientific culture these myths have a scientific basis. This expansion into a philosophical and religious viewpoint (or myth) of the whole has several logical steps or stages: (i) it begins with the understandable conviction (or faith) that science has no cognitive limits; proceeds (ii) to the more dubious and yet (in a scientific culture) plausible assertion that science represents the *only* cognitive touch with reality, that what science knows, and *only* what it knows, 'is the case' (e.g., instrumentalism, naturalism, and positivism); and it concludes (iii) with the metaphysical/theological proposition that what is ultimately real, and is therefore the source and origin of all else, is matter in motion, the blind physical processes of nature, which natural science has uncovered in its own limited inquiries. Thus by a movement from a methodological (heuristic) principle of inquiry into a metaphysical, substantive assertion about ultimate reality, we arrive at that philosophy, that 'atheistic and humanistic religious perspective' of which science as such is accused by the Creationists. That many in the scientific community have in fact participated in, witnessed to, and proclaimed as science this 'expansion' into a religious perspective, there can be little doubt. All too many distinguished scientists have stated that 'science now knows' that 'the religious theories of the past are outmoded myths' and 'the origins of all things lie in the material nature we find around us'.[6] In thus equating science with their own naturalistic humanism, these 'experts' are not acting at all as objective scientific minds; rather, they are expressing, quite without logical ground or empirical testing, the folk-wisdom or popular pieties abounding among the academic circles of an advanced scientific culture.

Although this identification of science with a non-theistic perspective, and of both with 'modern civilisation', is by no means shared by all scientists or teachers of science, nevertheless the assumption that religion represents early myths now outmoded by scientific developments pervades much of the writing and teaching of science. It should, moreover, be recognised that this widespread identification of religion with primitive myth, and of modernity and science with naturalistic humanism, has been one of the 'breeders' of the Creationist reaction. Each time a student in a high school comes home and says to his or her parents, 'Well, we learned today in science class that Genesis is wrong', two new converts to Creationism are created. In graduate programmes in science, complex questions of the relation of science and its truths to other aspects of culture: to politics, economics, morals, art, philosophy, and religion, have been blithely ignored. The history and philosophy of science—which *do* deal with these relations— are absent from most scientific programmes, located in the history and philosophy departments across the campus. Many scientists and teachers of science are quite unaware of modern, post-critical and post-scientific interpretations of religious traditions and beliefs—for example, of the Book of Genesis. Thus, most of them are utterly equipped to deal, however briefly, with the tricky interface of religious and scientific explanations of origins when questions concerning these relations arise, as they must in the teaching of science. The result is that all too often science is naïvely and perhaps innocently taught as replacing religion, rather than as supplementing, purifying, and informing it. Both the religious and the scientific communities bear real responsibility for the crisis represented by these laws.

(3) Modern scientific culture has taken many unexpected turns. Perhaps most surprising of all has been the reappearance of the religious in power and in pervasiveness in the midst of that culture. In the nineteenth and early twentieth centuries, it was widely assumed that religion and the religious were on the wane, made unnecessary by the growing power and influence of science, liberalism, and democracy; an age of scientific rationalism was replacing earlier ages dominated by religion. By the last decades of the twentieth century, this vision of a totally secular world, cleansed of the religious, seems an illusory fantasy indeed. The forms of the religious that have reappeared, moreover, do not represent so much the traditional religions of the West as new political or social religions (ideologies) on the one hand and non-Western traditions on the other. The public life of advanced cultures such as Germany and Italy, including their science and technology, have each found themselves dominated and directed by a fanatical and so, in the end, destructive, faith; Russia, much of Eastern Europe, and China have been gripped, transformed, and tightly controlled by another 'social religion' or ideology; and America itself, for all its boasted pragmatism, teeters now and again on the edge of dangerous ideological domination by the Right Wing. In the private life of modern Western people, via many traditional non-Western religions (Hindu, Buddhist, and Sikh) and some newly-appearing sects, the religious has been equally pervasive—clearly filling some deep-felt need, or needs, unsatisfied by scientific, technological, and industrial developments. And finally, among other social groups, fundamentalist and charismatic forms of traditional religions have appeared in new strength, forming, as we have noted, the religious background for the Creationist movement.

It is clear that the religious is a permanent, pervasive and always central aspect of corporate and individual life, generated out of fundamental human capacities and needs. Not only is there the ever-present necessity for a unified set of symbols giving a pattern to all of experience and so nature, history, the community, and the self in a coherent and meaningful version; without this, the self and its community have no identity, standards, tasks, or destiny, and little cultural or personal life is possible. Even more, at certain periods of social tension or breakdown, when assumed structures of social and personal security are badly shaken, deep anxieties arise, confidence lags or disappears, and the need for both a renewed certainty and a persistent courage waxes strong. To its own surprise, a technological age has found itself generating precisely such an especially religious period. For the anxieties, frustrations, and dilemmas of an advanced technological age are sharper, the menace of its social future more ominous, and so the sense of the continuing tragedy and estrangement of human existence deeper than at other times. Thus has the religious reappeared as ideology, as sect, and as dogmatism, not *despite* science and technology but precisely *because* of them. Here lies the deepest root of that 'union of science and the religious' spoken of early as the central quality of this controversy. In a scientific age, every form of religion becomes, as we noted, 'scientific'; even more in a technological age, many of those persons, in scientific professions and out of them, who live amidst the terrors as well as the benefits of technology, are impelled into religion in one of its many forms. And, as history continually illustrates, the religious (as do science and technology) appears in demonic, uncreative forms.

If, then, the evidence indicates that the religious is a permanent aspect both of human society and of personal existence, even in a scientific and technological culture, then immediately the issues involved with the relation of science to religion become of major significance to the social health of any community. These two aspects of culture *will* be related to one another in any case; possibly in the destructive form of the ideological domination by an irrational faith of the scientific and technological communities, as in Germany, in Russia, and in any Right-Wing America. Such a union is infinitely dangerous to both science and religion, as well as to the world at large. And we should note that in each of the cases mentioned above, this destructive union has been welcomed and supported as much, if not more, by the scientific community as a whole as by the religious communities involved. A religious community which allows its own dogmatic intolerance and its irresponsibility to the world to expand only aids in such false union. A scientific community that ignores the relation of its truth and its life to law, to morals, and to fundamental religious symbols also only makes itself and its culture vulnerable to ideological capitulation. Ignorance of the religious in both its demonic and its creative forms can be even more fatal for a scientific culture than ignorance of new scientific and technological developments. The implications of this controversy for our educational programme in the humanities, as well as in the sciences, are immense.

The questions how this interpretation understands *religious symbols* and how the latter may be related to the models, formulae, and theories of science (as of history and of philosophy) are too complex an issue for this brief document. Suffice it here to suggest that religious symbols concern the ultimate *horizon* of experience: its source and origin, its principles of fundamental reality and structure, its goal and destiny—in a word, its limits, its grounds, and its sources of ultimate hope. Every culture and every person must live within such a coherent, all-encompassing structure; from it derive the fundamental aims and norms of life; on it rests its most essential dependence, courage, and confidence. This ultimate horizon must be related to, as well as distinguished from, inquiry into the *foreground* of existence; into nature, history, society, and the self—in a word, the 'sciences'. As we noted, such a union of the religious and the scientific, of the ultimate base and the immediate character of life, will be there in any case. It is well that it is both rational and sane on the one hand, and responsible and humane on the other.

Notes

1. Almost all scientific commentators on the trial (see the articles in *Science*, in *Discovery*, and in the major engineering journals) failed to see this point. They understood the case between the ACLU and the Creationists to represent an argument over which one was science, and so which

one was true—and they saw our victory as arising because the scientific witnesses showed (as they indeed did) the scientific validity of evolutionary science and the scientific invalidity of Creation-Science. To them—if they even mentioned it—the religious testimony was, at best, 'an introduction' to the main thrust of the ACLU argument. This, of course, misinterpreted the whole legal situation. The scientific invalidity of Creation-Science is important but not central to their contravention of the First Amendment; what *was* important in that connection was that this model was shown to be religious and its establishment to represent the establishment of religion—and this only the witnesses from religious studies could effectively do.

2. In the long Christian, Jewish, Islamic traditions, of course, philosophers in each of these groups have sought to establish and define the concept of God 'philosophically'—the natural theology referred to earlier. Whether their philosophical efforts succeeded or not in reaching and describing 'the God of Abraham, Isaac, and Jacob' remains a subject of intense controversy. Nevertheless, even if natural theology be possible and relevant, still the concept thus philosophically (and so 'secularly' established) remains a *religious* concept, namely, 'God', the sole object of worship, obedience, and devotion of these traditions—as each of the theologians who did natural theology has emphasised. And if one looks cross-culturally at the results of these philosophical efforts, it is plain that the 'God' thereby established philosophically is, in comparison to the results of Hindu or Buddhist 'natural thology', clearly the God of the Christian, Jewish, or Islamic religious traditions and not of any other religious tradition.

3. When Senator James Holstead, the somewhat vacant legislator who introduced the bill in Arkansas, was asked how he, a person politically educated to know better, could have introduced a bill 'that favoured one kind of religion over others', he blandly replied: 'It does not favour any particular religion at all, since all religions agree on it. Why, the Methodists, the Baptists, even the Catholics, believe in Creation!' Since these 'widely different' groups represented, for Senator Holstead, *all* the variety of religions in present human existence, one can understand why he failed to see the favouritism involved in this law! (And he had a Vanderbilt BA.)

4. David Hume 'An Essay on Miracles' *An Essay on Human Understanding*.

5. See *Reaping the Whirlwind: A Christian Interpretation of History* (New York 1976).

6. See a new volume edited by C. Leon Harris, *Readings in Evolutionary Theory* (State University System of New York Press), in which the first section is entitled, 'Pre-Scientific Myths', with Genesis the main example, and the third chapter is entitled, 'The Infanticide of Science', with St Augustine and the middle ages the central example.

William Warthling

Pierre Teilhard de Chardin:
The Case Reopened

MANY THEOLOGIANS do not read science. Many scientists do not read theology. Many in both inquiries do not know that at some frontiers of both disciplines a truce has been declared. It is possible that they may be allies within a short time. If what Iris Murdoch says is true, that

> words constitute the ultimate texture and stuff of our moral being since they are the most refined and delicate and detailed, as well as the most universally used and understood, of the symbolisms whereby we express ourselves into existence[1]

then it is easy to understand why theologians might still be threatened by science. The theologian works at the word about God. The labour is to reach far into the self, all the years of accumulated experience and reflection, and bring out a word about self, world, God. They must necessarily be involved in their discourse, anxious about its outcome, needing support in the tenuous project. Their task is a work of webs, filaments, a gossamer project. From the theological horizon, science can look terribly exact. A precise vocabulary is aimed at measurable and definite targets. Autopsy, calipers are there to provide the data. Emotion and passion may seem far from scientific inquiry. Theologians may suspect that scientists would never understand theology. Nothing could be further from the truth. Scientists at the leading edge of the discipline share the *angst* and the wonder of the theologian.

Paul Dirac, winner of the Nobel Prize in Physics in 1933, who predicted the existence of anti-matter years before any form of it was observed, flew high and alone in austere scientific skies. Years later, sitting back, accomplished, he discussed the nature of scientific inquiry.

> It is more important to have beauty in one's equations than to have them fit experiment. It seems as though if one is working from a point of view of getting beauty in one's equations and if one has a really sound insight, one is on a sure line of progress.

How do you know when you have beauty there?

well you feel it. Just like beauty in a picture or beauty in music. You can't describe it, it's something—and if you don't feel it, you just have to accept that you're not susceptible to it. No one can explain it to you. If someone doesn't appreciate the beauty of music, what can you do? Give up on them! You just have to try and imagine what the universe is—like.[2]

Some scientists would be baffled at this description of science. They do not know that they too adhere to a metaphor, what the world seems like—currently. They often can see the devices they envision manufactured. The panoply of mechanical and technological products surrounds them and their work, providing a sort of soporific backdrop against which they see their progress. If nitrites which keep the meat looking pink cause cancer, progress is being made against cancer with chemotherapy. Within the paradigm things work.

Nor is theology exempt from the same kind of temptation. The socialisation of the metaphor can lull the theologian to become a mere technician of the sacred, and sleep. Socialisation of the myth into rituals, liturgies, and institutions confirms his or her enculturated choice of a confession of faith. The scandalous arguments among the different professions of Christianity, Islam, or Judaism have the strangest effect of making the discussion amongst them all the more intense and meaningful. Too few know that at the front a curious new liaison is developing.

Briefly, what the merger is all about is this. A game devised by Archibald Wheeler, a physicist at the University of Texas, initiates one quickly. A player is asked to leave the room to return later and find, through questioning, a word that all inside have chosen. While the first player is outside the room, those within change the rules. They may all answer 'yes' or 'no' as they please—provided they have a word in mind that fits both his own reply and all the previous replies. The victim returns and begins the probe. The activity that follows is like science. Nature yields answers according to the kinds of questions the seeker puts to it. The questioner presumes a word exists when the questioning begins, just as physicists presume in experiment that a certain reality exists. Ms Murdoch has warned that the word comes into being through the questions raised, the physical world emerges from the observations made. If the player asks different questions, different answers occur. No phenomenon is a phenomenon until it is observed.

The Phenomenon of Man[3] observed by Pierre Teilhard de Chardin is a remarkable work of observation. Is it naive? There seem to be no problems with mind and body, interiority and exteriority. If I may ask a question with Thomas Altizer, are there any dialectics present in Chardin?[4] After Kant, can anyone really write like this? There seems to be only one reality going forward—God. The charge of naiveté is not possible. For here we clearly find a first rate scientist, a profoundly religious man, often deeply spiritual. Poetry is followed by astute observation described with biological and anthropological acumen. Here and there are breathless lapses into mystical utterance. It is, perhaps, a small wonder that the Vatican thought that this might be pantheism. It is even smaller a wonder that many scientists eschew his theological discourse and that many theologians avoid his scientific jargon. He does not write from their horizons. The bias of both disciplines has cautiously placed him on the fringe of both their separate literatures. There, on the fringe, to be sure he is sold in abundance, in airport bookstores, mind body institutes, East-West store fronts. The confusion of our days has many individuals so desperate that they search, often with remarkably good instincts, for some last truth.[5] Many enthusiasts, 'true-believers', clutch their paperbacks with the tenacity of those pressed down to and, alas, limited to a single truth. This all the more confirms the bias of others. Teilhard's case must be reopened.

Thomas Kuhn's *The Structure of Scientific Revolutions*[6] and Earl MacCormac's

Metaphor and Myth in Science and Religion[7] have patiently laid bare the bones of the discussion. It seems that as the observer or theologian looks at what cosmos or God might be, all the creative powers of imagination and intellect are summoned to name what is seen. This act, of seeing and naming the all, God, is a limit-experience and can only be articulated in limit or peak language, metaphor, what things look like. The scientist as artist in this naming act sees it all, perhaps, as waves or as particles or, more recently, the new metaphor used is wavicles. Wavicles are an attempt to negotiate between the Fermi statistics which hold for reality seen as particles of reality, and Böse's statistics which hold for reality seen as waves of energy. Ian Barbour, with *Myth, Models and Paradigms*[8] again has driven the point home. In Barbour, the model or paradigm hold sway against newcomers. Bias and power favour the current metaphor. New models have to struggle to take hold in a particular community of inquiry. David Tracy has named the virtue needed most in both disciplines in his discussion of the scientific dimension of religion and the religious dimension of science.[9] The drive to that dimension, of course, is named self-transcendence. In this case, self-transcendence can translate as the elimination of any bias favouring one's own paradigm or academic province. More self-transcendence is needed than merely the temporary stay of execution that was forced on both communities recently. One instance of that quality of self-transcendence on the part of physicists was the strange discussion of black holes. In sheer exasperation science threw up its hands—how can one discuss, measure a nothing? The classical texts with chapter headings dealing with being and non-being were retrieved, dusted off. Strange bedfellows, the medieval scholastics and contemporary physicists. Yet another occasion of self-transcendence was the hypothesis of quarks and gluons in subatomic physics. Why that certain number of particles? Scientists now appeal to classical Aristotelian definitions of elegance and beauty, a harmonic balanced number as more likely to fit experiment.

In Chardin, the truce is already in effect. In the priest, the old obedience is there. He never published when proscribed from doing so. In his writings, the letters, what appears to be a profound and believable piety emerges from time to time. Enmeshed warp and woof in this is strict field-notebook science. One can imagine the condition of his boots clambering through rocky terrain, an ardent Jacques Costeau of the Mongolian wilderness. Not a shred of data escapes this Jean Henri Fabre of the skeletal, the fossilised. And so he wrote. It is all there. Rarely is there ontology, metaphysics. The upper case vision/metaphor/hypothesis which unifies the whole myriad of observation is God.

Things have a 'without' which is seen, measured, counted. This aspect is one with the 'within' which can only be intuited. In the context of Chardin's grand metaphor, radial energy is the within of things, tangential energy the without. Both are inextricably linked as aspects of one reality. The presence of 'withinness', radial energy, love pushing forward in the centre of cosmos is responsible for phyletic spiritualisation which is the tendency of matter where possible to strive toward the spiritual. It is precisely in these neologisms that we find really new hypothesis. He followed God as a hunter tracks as yet unnamed, already intuited beast. The depth of the tracks, the leavings of what is eaten, droppings, a skirmish recorded in the snow, provide the intuitions which make it possible to name the mystery—God. In a brave, completely new concept, 'pre-matter', a primordiality of chaotic energies emerges from a cataclysm of God's own fully actuated, exploratory Love. Total withinness burst into the without, the visible. What will be the world has happened. Alpha. Unimaginable eons needed for the formation of matter pass. Matter is energies related to one another in a regular configuration. The pressure is on. Everything goes. The emergence of the periodic table, that is, all the possible combinations of energy given the temperature and pressures of the place where matter is forming, cannot stop the search. Critical breakthroughs, leaps, because of 'temperature and pressure' (now used metaphorically) occur. Matter is now present, but that energy diffusing development in the cosmos is not enough to allay the built-in, forward drive. Megamolecules complexify into microorganism but in the fluctuating process are annihilated by an unfavourable temperature in a newly formed earth atmosphere.

And yet time, now that it exists because of the differentiation of energies into matter, is not an obex. In fact time, rather than the usually conceived of catalyst leading to disintegration or entropy, is the central factor in favour of the labyrinthine explorations nature needs in trying all so that something can succeed. Microorganism, no more dazzling a critical breakthrough than the appearance of matter, is yet another runner in the desperate relay race of cosmos toward an as yet undetermined end. Sphere upon sphere tentatively appear, fade, reappear to slowly take hold, and finally become firm: lithosphere, atmosphere, hydrosphere, biosphere are superimposed one upon the other. Teilhard dared to add the noosphere as the appearance of mind. This is dramatic, to be sure, but in fact is no more a marvel than the original cosmogony where love as energy became pre-matter. Here and there too, man who knows that he knows, appeared and faded, still too tenuous an attempt on the part of nature to be conscious of itself. Time again. A foothold, a favourable period and man is seated squarely and centredly in the always not yet decided thrust forward. Only one rule pervades: forward growth, Orthogenesis. Fire once discovered in man's early fumblings is now discovered again. The critical breakthrough to the appearance of love forever changes the footprint or pattern of cosmos, Prior to this final sphere of love, the Christosphere, differentiation, birfurcation, the disparate, the discrete were the way forward. One might think of a dandelion launching its frantic thousands of seeds, or the fruit tree with hundreds of thousands of blossoms, all the same message to be flung forward. In the case of man as well in the procreative act, some one one hundred and fifty million spermatozoa are sent onward to provide a single human being. Which blossom, which trout will succeed is a matter of the fluctuating Olympics of nature. Genetic drift, recession, environment are all in the fabric.

Now with the appearance of love a trend not seen before bends, coaxes all into a convergence clearly headed toward coordinates hauntingly familiar—Omega. Is it Alpha, the beginning again? No, one can never put one's foot into the river the same way again. The water has moved on. Even the foot is different, already having experienced river. God too is in process. Omega is God having experienced exploratory love—cosmos. Alpha was God before that. Words disclose the speaker. The Word-Christ entered and shook the fabric of the whole cosmic cloth. The artist shows up in the art. The infinite possibilities below are an expression of the infinity above. As above, so below. As God loved the world and this actuating love creates and is partner in the ongoing flux of cosmos, so too Christosphere, love in return for God, appears on earth and affects God's very self. Those who have loved know this. They are created by those whom they love and in turn create those whom they love. The appearance of love is the cosmic Christ. This second coming is transformed in Chardin from a literal, one time cinematic spectacle, into the twist imparted to the very strands of the fabric of the cosmos. Teilhard believed that love is now barely present, soon to become a final visible sphere, the final layer. The aggregate of individuals will form a superorganism, a collective combining the whole of human consciousness, now converging because of the appearance of love. The special properties of radial energy protect the fragility of the individual in the collective, that property being love. Theologians of every religion reflect on self, the world, and God and refer to the spirit of their scriptures and tradition in so doing. A particular passage, author, or theme in the given scripture often is favoured because of the support it lends the metaphor they are developing. This is true of Teilhard who finds in Paul:

F

from the beginning till now the entire creation, as we know, has been groaning in one great act of giving birth. . . .[10]

Unity in love is offered as a final end of all.

Now, for any strict empiricist who may be trembling at the thought of love starting anything, let alone cosmos, and for any theologian shuddering at the possibility of God actually being affected by our love, one might remember Hubbell's famous remark, 'the universe is not queerer than we suppose, it is queerer than we can suppose'. The imagination in art and science must be turned loose. Chardin wrote before the currently enforced scientific and theological dialogue was in effect. In him we have a man who was stunned by a vibrant vision of self, world, and God. *The Divine Milieu*[11] for those lacking a background in science, *The Phenomenon of Man*[12] for those prepared for scientific writing are still the best starting points. For those who, interested in the man, look for more biography, there is *The Making of a Mind: Letters from a Soldier-Priest.*[13] Poets will continue in *Building the Earth*[14] and *Hymn of the Universe.*[15] Where *The Phenomenon of Man* was the entry-point, further reading in *The Activation of Energy*[16] or *The Appearance of Man*[17] are where the physicist and the entropologist are invited to continue.

Chardin has followed Dirac's advice. Keeping beauty in the hypothesis, and going forward with a sound insight as a basis for his metaphor—he has tried to imagine what the universe is like. In its multiplicity, infinite possibilities, it is a reflection of the Infinite. In so far as cosmos strives toward love it is amazingly like God. Teilhard de Chardin's perusal of these metaphors on behalf of us all, scientists and theologians alike, is surely a rare achievement. Dare I suggest that within the confines of his language a science or theology uninformed by passion and love is doomed to be a side-issue in the ongoing conversation? Might it be further suggested, conversely, that when endeavours in both these disciplines are informed by love, the convergence peculiar to efforts in tune with the cosmos converging would end the problems brought before the reader in the beginning of this Bulletin?

Notes

1. Iris Murdoch interviewed in *Time Magazine*, 29 May 1972.
2. Paul Dirac as quoted by Horace Freeland Judson in an interview. *Newsweek*, 17 November 1980.
3. *Phenomenon of Man* (New York 1959).
4. Mircea Eliade and Thomas Altizer *The Dialectic of the Sacred and the Profane* (Philadelphia 1963). Chapter 6, 'Space and the Sacred'.
5. For those interested: the Teilhard Centre for the Future of Man, 23 Kensington Square, London W8 5HN, England. Bibliographies, articles available.
6. Thomas S. Kuhn *The Structure of Scientific Revolutions* (Chicago 1962).
7. Earl MacCormac *Metaphor and Myth in Science and Religion* (Durham, North Carolina 1976).
8. Ian Barbour *Myths, Models and Paradigms* (New York 1959).
9. David Tracy 'The Religious Dimension of Science' in *Concilium* 81 (January 1973), 128-135.
10. Romans 8:22.
11. *The Divine Milieu* (New York 1965).
12. *The Phenomenon of Man* (New York 1965).
13. *The Making of a Mind* (New York 1965).
14. *Building the Earth* (New York 1969).

15. *Hymn of the Universe* (New York 1969).
16. *Activation of Energy* (Helen and Kurt Wolff 1971).
17. *The Appearance of Man* (New York 1966).

Ursula King

Modern Cosmology and Eastern Thought, Especially Hinduism. Some brief bibliographical considerations

A CONTEMPORARY textbook on modern cosmology begins by saying that 'the exploration of the Universe, as conducted by physicists, astronomers and cosmologists, is one of the greatest intellectual adventures of the mid-twentieth century. It is no exaggeration to say that their achievements, especially in the last few years, constitute a revolution in our knowledge and understanding of the Universe without parallel in the whole recorded history of mankind' (D. W. Sciama *Modern Cosmology* (Cambridge 1975) p. vii). The excitement of modern physical and cosmological theories, whether in relation to the understanding of the nature of matter, energy, light, space and time, or the history and structure of the universe as a whole and the place of consciousness within it, has led some scientists to a new interest in philosophical speculations about the origin and nature of the universe. Although still a minority, a growing number of writers consider certain aspects of traditional Eastern thought as particularly illuminating for the newly emerging world view based on the contemporary study of physics and astrophysics. Some see astounding parallels in Eastern religions, others speak even of a convergence of Eastern mysticism and the new physics. Correspondences are alleged to exist in Indian and Chinese thought, especially in Taoism, but also in specific teachings of Hinduism and Buddhism, for example Tantra, Zen or Madhyamika Buddhism. One of the best-known explorations into these parallelisms is F. Capra *The Tao of Physics. An Exploration of the Parallels between Modern Physics and Eastern Mysticism* (London 1975). Many other authors, perhaps more often popular writers on science rather than scientists themselves, have referred to affinities between modern cosmological thinking and ancient Eastern mystical teachings (among others see G. Zakov *The Dancing Wu Li Masters. An Overview of the New Physics* (New York 1979) which emphasises the parallelism to Chinese thought, whereas M. Talbot's *Mysticism and the New Physics* (London 1981) primarily thinks of Indian thought when he stresses the 'confluence of mysticism and physics'). The most frequently mentioned parallels are the Eastern view of the interconnected unity of all things, of an unbroken wholeness recovered in certain experiences of meditation (brief allusion to this is made in the very sensitively written work by D. Bohm *Wholeness and the Implicate Order* (London 1980)), a unity which transcends all opposites and is grounded in a totally different

76

experience of space-time. Another central aspect is the dynamic nature of the universe, often seen as symbolised in the dance of creation of the great Indian god Siva. Parallels are also drawn with Buddhist speculations on emptiness and form (see Capra and Talbot, the works cited), the interpretation of different levels of the universe and Hindu views of panentheism as well as other Indian speculations concerning the relationship between matter and mind, the nature of consciousness and reality as a whole. Here, parallels are sometimes indicated with Indian Advaita Vedanta teachings on the oneness of *atam* and *brahman* (see the two articles by Jantsch in *Evolution and Consciousness. Human Systems in Transition* eds. E. Jantsch and C. H. Waddington (Reading/Mass. 1976)) and also the ultimate unreality of the cosmos as *maya*.

What foundations are there in traditional Eastern thought for these parallels with modern cosmological and physical thinking alluded to by Western writers on science? The world view and reality-structure emerging from modern nuclear, quantum and astrophysics is truly dazzling and yet when it comes to actual comparisons with Eastern thought, the parallels seem sometimes far-fetched, sometimes ill-founded and most often characterised by an eclectic vagueness and rash generalisations which scientists would not wish to apply to their own field of specialisation. The articulation of the modern cosmological world view is far more precise in its scientific details than in its references to Eastern thought where little attention is paid to the complexity of the many philosophical and religious traditions of the East. The historical and experimental multidimensionality of the different Eastern mysticism as well as the existence of a rich tradition of Western mysticisms are usually totally ignored in order to present a facile coherence of 'the Eastern way of mysticism' and the world view now emerging from modern science.

There exist of course many interesting and thought-provoking similar elements, correspondences and parallel insights in the different religious and philosophical traditions of the East and these deserve closest considerations. Yet the whole subject is beset by a lack of clarity and consensus. Whilst 'cosmology' may have a precisely defined meaning in modern science, the term is often used more loosely in religious and philosophical thought. Thus any attempt to compare modern cosmology with Eastern thought is bound to come across fuzzy and confused areas as different people will base their comparisons on very different premises. Besides the lack of a clear definition of what 'cosmology' includes, further problems arise as material from the East dealing with these correspondences is limited and difficult of access, and whatever material there is, is often vague and imprecise. These brief bibliographical notes have the modest aim of pointing to where a few elements for more precise information can be gathered, but they cannot themselves offer a contribution to the necessary comparisons between modern cosmology and Eastern thought which would be better grounded in its details than those undertaken so far.

Many parallels are found in Indian thought, especially Hinduism, and I shall limit my discussion to this tradition. The difficulty, however, is that no full-scale study exists as yet where these parallels are systematically and critically examined. Contemporary Indian philosophers seem to show little inclination for a systematic exploration of this subject; indologists rarely look at Indian scientific materials; writers on the philosophy of science by and large restrict themselves to established areas of the Western intellectual tradition, and even a journal such as *Zygon*, explicitly devoted to the encounter between science and religion, indicates over the more than fifteen years of its existence that the modern scientific world view has only been reflected on from the perspective of Western philosophical and theological resources. Its index includes no references to non-Western materials, whether drawn from the Indian, Chinese, African or Islamic heritage. We urgently need a more inclusive framework which can incorporate elements from different philosophical and religious traditions and create a truly integral synthesis. Some Indian thinkers, such as Sri Aurobindo and S. Radhakrishnan, are said to have created a synthesis between Eastern and Western ways of thinking, but in spite of such a claim one can maintain that the mutual interpenetration of Western and Eastern thought in the area of religion and science has hardly begun. There is also the danger of a false reification of religious systems such as 'Hinduism' which, as an abstraction, can hardly initiate an encounter with the modern scientific world view. On the contrary, the transposition of traditional thought into dynamic categories will always depend on the creativity and commitment of particular individuals who know the inherent strength of their faith as well as the achievements of modern science. At present we know of no Hindu work which would parallel various Western attempts to combine a theological understanding of creation with the insights of modern cosmology, as found for example in the discussion of A. R. Peacocke *Creation and the World of Science* (Oxford 1979; see *ibid.*: 360-363 'Modern Atomic Physics and Eastern Mystical Thought' for a discussion of F. Capra, the work cited earlier). But we possess a historical survey of the attitudes of Indian scientists to traditional religion and the views of Hindu reformers on modern science in D. L. Gosling, *Science and Religion in India* (Madras 1976). This includes questions regarding reactions to theories about the origin of the universe and biological evolution which, some scientists admitted, were a cause of conflict for their understanding of traditional Hinduism.

Those who wish to explore the parallels in Indian thought further than the brief allusions in popular Western writings can find more information in the following. I shall first list works dealing with traditional Indian cosmology, then I shall mention some of the main themes most often alluded to in discussing parallels between Indian and modern scientific thought.

1. TRADITIONAL INDIAN COSMOLOGY

A succinct survey of Indian cosmological thought is provided by R. F. Gombrich's article 'Ancient Indian Cosmology' (in *Ancient Cosmologies*, eds. C. Blacker and M. Loewe (London 1975) pp. 110-142). It highlights the complexity of Indian traditions and distinguishes between four different cosmologies: the early Vedic cosmology is very different from the later Hindu, Buddhist and Jain cosmologies which arose after about 500 BC. In all the hymns of the *Rig Veda*, the earliest Indian document dating from the second half of the second millenium BC, numerous speculations about the nature and origin of the universe can be traced. Many profound cosmological concepts are embedded in the ancient Vedic myths. A detailed analysis of these is given in 'Theories of Creation in the *Rig Veda*' by W. Norman Brown (*Journal of the American Oriental Society* 85, 1965, 23-34). For further philosophical discussions of the rich cosmological speculations in Vedic literature consult J. N. Chakravarty *The Concept of Cosmic Harmony in the Rig Veda* (Mysore 1966) and R. N. Dandekar 'Universe in Vedic Thought' (in *India Maior* eds. J. Ensink and P. Jaeffke (Leiden 1972) pp. 91-114) which deals with Vedic thinking on cosmogony, cosmography, mythology, ethics and eschatology. Many more details about the nature of the Vedic cosmos and the cosmic sacrifice are discussed in A. Danielou *Hindu Polytheism* (New York 1964) and the richest contemporary meditation on the wholeness of the Vedic world view is contained in R. Panikkar *The Vedic Experience* (London 1977).

Gombrich discusses the Buddhist and Jain notions of space and time and the cyclical nature of the universe, found in classical Hindu cosmology, also referred to as Puranic cosmology (found in different texts called *Puranas*, mostly composed in the first millenium AD). The most detailed analysis of the medieval Puranic cosmogonies and their powerful visions of the periodic creation, absorption and recreation of the universe is found in the masterly studies by M. Biardeau 'Etudes de mythologie Hindoue— cosmogonies puraniques' (in *Bulletin de l'École Francaise d'Extrême Orient* 54, 1968, 19-45; 55, 1969, 59-105; 58, 1971, 17-89).

Anyone interested in precise information about possible parallels in earlier Indian scientific thought would do well to consult *A Concise History of Science in India* (edited by D. M. Bose, S. N. Sen, B. V. Subbarayappa, Indian National Science Academy, New Delhi 1971). Not only does it contain a survey of the different literary sources available to us (whether Vedic, Hindu, Buddhist or Jain) but it also includes a detailed article on 'Astronomy' by S. N. Sen (pp. 58-135), particularly interesting for the topics found in traditional astronomical texts, especially in the influential *Surya Siddhanta* (about 4th-12th century AD) concerned with sophisticated mathematical speculations about the different ages of the cyclical universe, stellar configurations and divisions of time. Hindu stellar configurations have been compared with modern star catalogues and historians of Hindu astronomy have found some amazing parallels to modern findings. The interrelationship between Indian, Greek, Chinese, and Arabic astronomy is also briefly discussed. Another article in the same volume, written by B. V. Subbarayappa, deals with 'The Physical World: Views and Concepts' (pp. 445-483) which brings together the various Indian notions concerning matter and motion, causation, ideas about cosmic light, cosmic order and the regulating nature of the Vedic sacrifice, the doctrine of five elements, traditional Indian teachings on atomism, substances and their attributes, space and time, heat and light, and the phenomenon of sound, so important in philosophical speculations. Further philosophical elucidation of traditional scientific thought is found in the *Indian Journal of the History of Science* 4, 1969 (see the articles by T. M. P. Mahadevan 'Philosophical Trends *v.* History of Sciences of India— Orthodox Systems' pp. 27-41, which discusses theories of perception, causality and evolution, and G. C. Pande, 'Philosophical Trends and the History of Sciences in India—Heterodox Trends', *ibid* pp. 42-51, which looks at Buddhist and Jain theories from the standpoint of their relevance to scientific progress).

Another treatment of traditional cosmological thinking is found in the study by N. Bhattacharyya *History of Indian Cosmogonical Ideas* (New Delhi 1971). So far I have only been able to find one attempt to go beyond the mere historical exposition of traditional Indian cosmology and relate the latter to modern cosmological discoveries. In a simple, introductory way this is done by S. S. Bhattacharjee *The Hindu Theory of Cosmology. An Introduction to the Hindu View of Man and His Universe* (Calcutta 1978). The author underlines the complexity of Hindu cosmology and the difficulty of extracting early cosmogony from a maze of myths and metaphysics. He equally stresses the difficulty inherent in the Western indological approach to Indian thought, more usually concerned with literary rather than scientific knowledge about man and the universe. After considering the different creation myths, he deals with the *avatar* series of divine incarnations (also part of Puranic mythology) in an ascending line from animal to human form and reflects, as other Indians do, 'as to how it all became possible to offer a picture of evolution of the living form with such parallelism with the Darwinian theory of evolution' (p. 30). The Hindu view of the dissolution of the universe, called *pralaya*, is dealt with in great detail) especially the difference between the partial dissolution (after a specific time) and the complete, full-scale dissolution whereby the phenomenal universe reverts back to the latent pre-creation condition. The destruction of the universe proceeds first by fire and then by a cosmic flood. The author describes this *pralaya* as an 'anticreation process' and says that modern cosmology theoretically predicts something very similar to the Puranic accounts, indicating even a closeness in their respective calculations.

The universe is seen as moving through vast cosmic cycles. There is a dialectical dynamism in the cyclical rotation of the creation and dissolution processes which can be seen as spontaneous without the intervention of a conscious will. Many speculations in the *Puranas* are devoted to the calculations of the age of the universe and the different units of time. For comparison contemporary cosmological views are quoted which consider the deceleration of the galaxies as an indication that the universe will eventually collapse and that it may be forever cyclical, successively evolving and collapsing. According to a Western astronomer, some of the ancient Indian astronomical books, particularly the already mentioned *Surya Siddhanta*, 'appear to contain much knowledge, some of which is being empirically rediscovered and confirmed by modern technics, albeit slowly. This fact, though unrecognised by the world of astrophysics, will become more and more apparent' (p. 144).

There can be no doubt that some amazing Eastern parallels and insights exist which as yet await further exploration. They must not only be sought in the general areas of religion, philosophy and mysticism but also in specific scientific insights. The barriers for perpetuating our ignorance are many, however, linguistic, historical, philosophical, cultural and social. Yet if modern scientific knowledge aims to be truly comprehensive and if we hope to build a global, universal civilisation, as we must if we do not want to perish, it will be to our peril if we further ignore this fertile heritage from the East.

So far, there is no unified attempt at integrating these parallels into a greater vision and deeper understanding. I sought in vain for such lines of thinking in the comments provided for the exhibition on 'Indian Science' during the Festival of India (London 1982). Here, as elsewhere, the fascinating historical data were presented in an expository, descriptive manner without being consciously related to the perspective of modern scientific knowledge. Such an overall integration of ancient and modern Indian thought with current cosmological and physical theories is rarely ever attempted yet there are a number of subjects which frequently lend themselves to comparisons and parallels. I can only mention a few of these here.

2. SOME PARALLEL THEMES IN HINDU AND MODERN SCIENTIFIC THOUGHT

Evolution: Understood in its widest sense as a vast cosmic process of which the development of human beings is an integral part, the theory of evolution has a profound effect on religious self-understanding. The modern theory of evolution has perhaps attracted the most attention in India, whereas Western writers do not seem to look for Indian parallels here. In India it is particularly the great thinker Sri Aurobindo who has tried to combine a systematic integration of evolutionary thinking with a modern reinterpretation of Hinduism. His vast corpus of writings is available in the thirty volumes of the *Sri Aurobindo Centenary Library* (Pondicherry 1972). For further studies consult H. K. Kaul *Sri Aurobindo: A Descriptive Bibliography* (New Delhi 1972). A critical analysis of Sri Aurobindo's major works is found in *Six Pillars* ed. R. A. McDermott (New York 1974); see also Satprem *Sri Aurobindo or the Adventure of Consciousness* (New York 1968); *The Integral Philosophy of Sri Aurobindo* eds. H. Chaudhuri and F. Spiegelberg (London 1960).

Sri Aurobindo's application of evolutionary thinking to the reinterpretation of Hinduism is often compared with Teilhard de Chardin's attempt to rethink Christian doctrine in evolutionary categories. A comparison between the two is found in R. C. Zaehner *Evolution in Religion. A Study in Sri Aurobindo and Pierre Teilhard de Chardin* (Oxford 1971) and, taken from a much wider perspective, in B. Bruteau *Evolution toward Divinity. Teilhard de Chardin and the Hindu Traditions* (Wheaton/ USA 1974). However, Sri Aurobindo's use of evolution has not remained without its critics. In the opinion of J. Feys (*The Philosophy of Evolution in Sri Aurobindo and*

Teilhard de Chardin (Calcutta 1973)) who has undertaken a detailed philosophical analysis of Sri Auronbindo's works, the latter's understanding of evolution remains close to traditional emanationist thinking, as he understands evolution primarily as the gradual manifestation of what always existed; the Absolute, the divine life, first involves itself in matter before evolving out of it. There is nothing really new, no new being and no real growth. For other criticisms consult the 'Proceedings of the Seminar on Sri Aurobindo and the Concept of Evolution', *Indian Philosophical Annual* 8, 1972 (see especially C. T. K. Chari 'Some Issues about Sri Aurobindo's Evolutionism and Modern Knowledge' pp. 20-26).

With regard to evolutionary parallels in traditional Indian thought the *avatar* series is often pointed to, as already mentioned. For fuller information on the meaning of *avatar* see G. Parrinder *Avatar and Incarnation* (London 1970).

Dance of Siva: This is a theme which seems to have caught the imagination of Western scientists more than any other, a theme largely known from South Indian sculpture but only one of the many aspects of the multifaceted god Siva (the richness of which is documented in the detailed study on Siva mythology by W. D. O'Flaherty *Siva. The Erotic Ascetic* (Oxford 1973) where the dancing Siva does not figure at all). It is difficult to assess whether this exclusive interest in Siva's cosmic dance or 'Siva Nataraja' is primarily due to A. K. Coomaraswamy's influential essay *The Dance of Shiva* (London 1958). F. Capra's *The Tao of Physics* and other works depict the dancing Siva on the bookcover but make only the briefest reference to him (a more detailed discussion is found in A. Peacocke, in the work cited above, pp. 106-111, drawing largely on Coomaraswamy). Coomaraswamy's own source of inspiration was, besides sculpture, the fervent devotional hymns of South Indian Saivism but these are generally little known in the West. After reading about these in the works of René Grousset in the late 1940s, Teilhard de Chardin was already drawn to the vision of Siva's cosmic dance whose ambivalence as source of creation and destruction appealed to him. In Siva the forces of joy, suffering and energy pulsating through the universe come together. Teilhard, who saw the cosmos as animated by divine energy, felt that this cosmic dimension, symbolised by Siva's dance, has never been fully explored by Christianity (for a brief discussion of Teilhard's diary entries on 'Christ-Omega/Siva' see U. King *Towards a New Mysticism. Teilhard de Chardin and Eastern Religions* (London 1980) pp. 98).

Consciousness, pure awareness, meditation techniques and maya: Since the earliest times Indian philosophical speculation, beginning with the Upanishads, has reflected on the nature of consciousness and distinguished its different stages culminating in pure contentless awareness reached by specific meditation techniques. Later this was systematised in the non-dualist teachings of *Advaita Vedanta*, particularly by Shankara (9th century AD) and his school, grounded in a very sophisticated theory of epistemology and metaphysics. It is also linked to *maya vada* or the teaching that man's world-consciousness is ultimately illusionary. Much has been written on *maya* as well as on Indian monism, an influential feature of philosophical and religious thought, but one must not forget that there are many other schools of thought, too. However, when mentioning parallels between modern cosmological and Hindu thought, Western scientists frequently refer to ideas from Advaita Vedanta, but only in a general way and often taken out of context.

For an analysis of the Indian understanding of world-consciousness, self-consciousness and cosmic consciousness in relation to modern philosophical thinking see R. A. Sinari *The Structure of Indian Thought* (Springfield/Ill. 1970, esp. ch. 8 'The Phenomenology of Maya'). Other aspects are explored in the 'Proceedings of the

Seminar on the Concept of Consciousness', (*Ind. Phil. Ann.* vol. XI, 1976) where C. T. K. Chari's contribution on 'Quantum Mechanics and Concepts of Consciousness' (pp. 50-56) is of particular interest. He mentions that a number of Indian thinkers have taken the latest relativistic and quantum field theories as proofs for the Advaita teachings on *maya* and the formless *brahman*. Chari argues against both Indian philosophers and 'mystagogues in the West' who hail modern physicists as a kind of open sesame for an idealistic and mystical view of reality. He concludes that, on one hand, 'Advaita Vedanta (or any Indian metaphysical system for that matter) does not require official endorsement by quantum mechanics' and on the other hand he also warns that 'The glamour of the new physicist must not lead to a metaphysical abuse of it' (at p. 55). In contrast to this, the converging insights between Advaita Vedanta and modern physics are explored by the physicist C. F. von Weizsäcker, 'Who is the Knower in Physics?' and Queen Frederica of the Hellenes 'To Advaita Through Nuclear Physics' in the volume edited by T. M. P. Mahadevan *Spiritual Perspectives. Essays in Mysticism and Metaphysics* (New Delhi 1975, pp. 147-161; 162-170).

A far more detailed investigation into the relationship between meditation, states of consciousness and quantum mechanics is found in the scientific research on the transcendental meditation programme carried out by Maharishi Mahesh Yogi and his disciples. The Maharishi also points to the remarkable parallels between the picture of the universe developed by modern physics and his own teachings on the fundamental structure of subjective life, supported by Vedic cosmogony. Thus, his 'science of consciousness' is presented as a holistic integration of the biological, psychological and physical sciences. These claims deserve much closer critical examination than is possible here. For further details see Maharishi Mahesh Yogi *Science of Creative Intelligence: Knowledge and Experience* (Maharishi International University, Fuiggi Fonte/Italy, p. 172) and *Scientific Research on the Transcendental Meditation Program: Collected Papers 1* eds. D. W. Orme-Johnson and J. T. Farrow (1976).

Those who wish to explore further aspects of *maya* should consult the 'Seminar on the Concept of Maya' (*Ind. Phil. Ann.* 2 (1966) (Part II) and the article by N. A. Nikam 'The Problem of Creation: Concepts of Maya and Lila' (in H. Chaudhuri and F. Spiegelberg, the work cited above).

Karma and rebirth: The cyclical view of the universe is closely related to the teaching on *karma* and rebirth, occasionally mentioned in Western discussions but perhaps more of a stumbling block than help in drawing out parallel thoughts. However, the doctrine of *karma* is unintelligible without Indian cosmology as is clearly brought out in P. Bowes 'Morality and Cosmology in Hinduism' (*Theoria to Theory* 12 (1978) 97-109) and some of the problems arising out of modern genetics for the understanding of *karma* are alluded to in the discussion on 'Spirit and Science in India today' (*Theoria to Theory* 14 (1981) 273-289). A fuller philosophical treatment of *karma* and rebirth is found in the 'Proceedings of the Seminar on Karma and Rebirth' (*Ind. Phil. Ann.* 1, (1965); see also R. de Smet 'The Law of Karma: A Critical Examination' *ibid* 2 (1966) 328-335, and J. Hick 'The Idea of Rebirth—A Western Approach' 6 (1971) 89-101).

Modern physical and cosmological theories are characterised by an extreme openness and by a holistic, process-oriented thinking in which a teleological orientation is gaining new ground. The particular attraction of Indian thought, especially of Advaita Vedanta, for Western scientists, lies perhaps less in specific comparisons than in its general orientation well summarised by E. Deutsch when he states that Vedantic theory, 'like most of Indian thought subscribes to an emanationist theory that sees 'creation' as a natural unfolding of spirit in the world—and that it also sees, accordingly, that everything in nature, when properly seen, is invested with intrinsic spiritual worth.

Nature thus becomes infused with quality, with spiritual life, as well as with quantity, with measurable processes . . . The core ideas . . . are that nature is interconnected, that there is a continuity between man and all other living things, and that everything in nature, being an inseparable part of the whole and retaining its qualitative origin, has intrinsic value' ('Vedanta and Ecology', *Ind. Phil. Ann.* 6 (1971) 83). However, what this author applies to ecology is equally true for the new physics and cosmology, namely, that we should not turn to Vedanta for imitation or direct instruction but rather for inspiration and enriched understanding. If closer attention were paid to the varied teachings of different Vedanta schools, the qualified non-dualism of Ramanuja (11th century AD) and his thought about the world being the body of God would provide further ground for exploring parallels so far unheeded. Similarly, Arjuna's cosmic vision of God in his universal form or *visva rupa*, found in the *Bhagavad Gita* and other Indian texts, has not yet been given the attention it deserves for drawing out similar insights present in ancient religious and modern scientific thought. The divine energy of *shakti* (understood as a female force) and the deep insights expressed in the polar symbolism of Tantra also deserve closer attention that the mere mention in Western books. But this is equally true of Zen, Taoism and Buddhism.

As is evident from this brief discussion, the greatest difficulty in drawing out parallels lies in the lack of comparative studies of sufficient detail and quality regarding Eastern thought and modern cosmology. Only a vigorous support for the study of Eastern thought in all areas, whether in religion, philosophy, science or mysticism, will overcome the present ignorance and lack of discrimination. Although Western religion may have lost much of its traditional plausibility structure and a too narrowly conceived science may no longer still our spiritual hunger, we must be careful not to be taken in too easily by a new visionary physics and cosmology whose rash extrapolations regarding Eastern parallels may be more an indication of Western gullibility than give us a true picture of the rich reserves in the powers of wholeness and integration found in the religions and mysticisms of the East.

If the exploration of the universe and outer space is truly one of the greatest intellectual adventures today, we must ensure that we apply equally close attention and similarly demanding criteria to the exploration of inner space and the rich traditions of mystical insight that are part of our common human heritage, whether from East or West. Paying mere lip service to brief references and parallels simply will not do. We require an in-depth discipline (a word of some importance in both science and religion), exacting inquiries and an integration of scientific and mystical insights which truly match each other.

For additional information readers may like to consult the following publications which came to my notice since writing this article: Sal P. Restivo 'Parallels and Paradoxes in Modern Physics and Eastern Mysticism' in *Social Studies of Science* 8 (1978) 143-181; Tarthang Tulku *Dimensions of Thought: Current Explorations in Time, Space and Knowledge*, 2 vols., Dharma Publishing (Emeryville/California 1980). The teachings of Tarthang Tulku are followed by the 'Time, Space and Knowledge' group in Berkeley which now publishes a regular journal dealing with related matters discussed in this article.

See also R. H. Jones *Mysticism and Science: A Comparative Study of the Claims about Reality in Western Natural Science, Theravada Buddhism, and Advaita Vedanta*, PhD Thesis (Columbia University 1980).

PART III

Concluding Editorial Reflections

David Tracy, Nicholas Lash

Editorial Reflections

1. THE CHANGED SITUATION IN THEORY AND PRAXIS

A RETURN to the issues of cosmology seems both desirable and necessary in our present theological situation. That return is desirable largely because of significant shifts in the methods and contents of both theology and science. The return is necessary for two related reasons. First, there is a growing sense (occasioned by the ecological crisis and the threat of nuclear holocaust) that the 'anthropocentric' character of much contemporary theology must be challenged. Second, redemption itself cannot be understood without a relationship to creation; history cannot be understood without nature; the central categories of God and the self (and, therefore, society and history) cannot be fully grasped without reference to the category 'cosmos' or 'world'. Our present editorial reflection makes no claims to resolving this complex set of issues. The editors will be content if readers come to share their sense of how urgent cosmological concerns should be for all theology. Above all, it is the new 'status quaestionis' that must be understood before the new constructive theological work already begun in this issue and elsewhere can be assessed.

Our contemporary theological situation, to repeat, has changed in both theory and praxis. In both cases, the urgency of a theological return to cosmological interests seems clear. Consider, first, the situation from the point of view of contemporary theory. We forget all too easily that when theologians of early modernity (seventeenth century to early twentieth century) approached the relationships between science and theology the situation was, at best, deeply troubling. The scientific revolution was an intellectual event, as Butterfield remarks, which became a genuine intellectual revolution demanding a new 'thinking cap' in all disciplines. Indeed so radical was this intellectual revolution that Butterfield does not exaggerate when he states that, in comparison, even the Renaissance and the Reformation look like 'family quarrels'. There can be little doubt that the events symbolised by the names Copernicus, Galileo, Newton and Darwin changed forever the landscape of theological thought.

Neither the earlier 'warfare' between science and religion nor earlier 'concordist' proposals of theologians and philosophers in that long period of early modernity strike contemporaries as genuine options for our radically changed situation. The principal reasons for the collapse of both the confrontational ('warfare') model and the concordist model are to be found in the changed understandings of the content and the methods of both theology and science. For however alive earlier scientific models of mechanism, materialism, and positivism may be in the popular imagination (including the popular imagination of some scientists), the fact is that science itself has challenged the intellectual pretensions of these earlier models. It is not possible in an age where the content of science has been radically changed by evolutionary theories, relativity, quantum mechanics, the principle of indeterminacy, quarks, DNA research etc. simply to appeal to earlier mechanist or materialist models. The content of the sciences now disallows all easy appeals to these positions. In an ironic symbol of this momentous shift, we may say that science itself no longer has need of Laplace's materialist and mechanist hypothesis! The methods of science, moreover—as the articles by M. Hesse and L. Gilkey show—have themselves yielded to more modest self-appraisals. The recognition that reason too has a history has yielded, in the history of science, to proposals like those of T. Kuhn, S. Toulmin, M. Hesse and others that earlier positivist models for scientific self-understanding cannot survive the study of the history of science nor philosophical scrutiny by philosophers of science.

This shift in both the content and the self-understanding of the methods of science has occasioned, therefore, a new intellectual situation where the relationship of science and theology seems at once more promising and more difficult. It is more difficult in the properly theoretical sense that the issues are now far more complex and often highly technical: for this reason alone earlier models of sheer confrontation or easy concordism are inappropriate. It is more promising because the collapse of earlier mechanistic, materialist and positivist models has freed science itself to a sense of the ultimate mystery of reality and to a chastened but real willingness to dialogue with any plausible philosophical and theological cosmological hypotheses. The dialogue with process philosophers and theologians (including, in the wider sense of process thought, the followers of Teilhard de Chardin) is merely one well-known instance of this increasingly fruitful relationship.

It is equally important to recall, of course, that theological self-understanding and theological content have undergone analogous paradigm-shifts in the same period. Except for fundamentalists, earlier theological models based on a-historical and authoritarian understandings of theological claims have collapsed. Except for naive concordists, the technical complexities and tentative, hypothetical character of the theologically relevant issues in contemporary science do not yield themselves to easy solutions. Indeed, all easy solutions—scientific and theological—have been unmasked as, in effect, ideologies for intellectual élites functioning to assure the *status quo*.

Theologians, in sum, have learned their own form of chastened methodological and material modesty. Most now realise, as E. Schillebeeckx has observed, that theology is an interpretive enterprise that attempts to establish 'mutually critical correlations' (in both theory and praxis) between interpretations of our contemporary situation and interpretations of the Christian tradition. The rise of historical consciousness in the nineteenth century and the radical sense of historicity in the early twentieth century have impelled modern theology, as B. Lonergan has observed, to abandon its commitments to 'classical consciousness' (including classical cosmologies). Granted these gains, one still cannot avoid the impression that (with some notable exceptions) contemporary theology, although strong on interpretations of history and redemption, is relatively weak on interpretations of nature and creation. Part of the reason for this is undoubtedly that earlier 'warfares' between science and theology and the collapse of earlier concordist proposals have rendered most theologians reluctant to re-enter the dialogue with natural science. The fact that most theologians, by both predilection and training, find their most natural 'conversation-partners' in the human sciences has also served to encourage this same development. An allied fact is that the major gain of contemporary theology is the 'de-privatising' of theology. This de-privatising has been best developed by political and liberation theologies faithful to a practical reason cognisant of massive global suffering. Yet those very same theologies can often encourage a relative lack of interest in strictly cosmological questions.

It should now be clear, especially to readers of *Concilium*, that a major accomplishment and a still major need of contemporary theology is to work out political and liberation theologies faithful to the demands of both theory and praxis in our grave historical situation. These theologies correctly retrieve the resources of the Christian tradition to help transform that situation of massive global suffering. Their critical analysis of the complicity in 'privatisation' on the part of earlier modern existentialist and transcendental theologies can only be considered a gain for all theology. The suspicion that 'cosmological' interests in theology *can* function as a distraction from these historical responsibilities or even as ideology for intellectual élites (as in concordiat proposals) seems, admittedly, in order. The importance of these suspicions should not be domesticated. Indeed, any pure models of 'progress' unable and unwilling to face the tragedy and suffering in human existence fully deserve Christian theological suspicion.

Yet another suspicion must also inform our contemporary theological consciousness: is it possible for theology to be faithful to the demands of either our situation or the Christian tradition while continuing to ignore cosmological concerns? The question is not merely rhetorical. For the fact is that the rediscovery of 'history' by contemporary theology has not been matched by a parallel rediscovery of 'nature'. A well-nigh exclusive focus on the doctrine of redemption (as related to liberation and emancipation) has not been paralleled by new explorations of the doctrine of creation. Contemporary theology is in danger of developing interpretations of God and self (including the social self in society and history) while quietly dropping the traditional third category of 'world' or 'cosmos'. And yet—as the historical articles in this issue make clear—this absence would represent a major impoverishment of the Christian tradition and possibly a major distortion of Christian understandings of salvation and history by themselves.

It is, of course, necessary to continue to demythologise and de-ideologise traditional Christian theological understandings of 'world' or 'cosmos'. Nevertheless it is also crucial to realise that from the New Testament period (J. Collins) through the patristic (H. Chadwick) and the medieval periods (O. Pedersen) the same insight has prevailed: in J. Collins's forthright statement 'Human salvation cannot be divorced from our understanding of the world around us. The creation, too, is groaning in travail. . . . It is important . . . that we find a way to integrate human values with some cosmological understanding if our theology is to represent more than a fragment of existence'.

The Christian tradition needs the doctrine of creation even to understand fully its own doctrine of redemption (J. Buchanan). Contemporary Christian theology needs to recover a theology of nature—even to develop an adequate theology of history. No Christian theology can claim adequacy to the Christian tradition by, in effect, retrieving only God and the self (including the social and historical self) while quietly dropping 'world' out of the picture. The questions of cosmology are not properly understood as *only* concerned with the origin and natural structure of the world. Those cosmological questions include the destiny of the world, as well—including the destiny of human beings, indeed of history itself—as 'inextricably bound up' with the destiny of the cosmos. The deprivatising of theology has meant a return to real history (not merely historicity). But that return to history, in both theory and praxis, must also mean a return to nature.

These more theoretical questions, moreover, are intensified if we note two central practical concerns of our contemporary period. Those two grave issues—in summary form, the ecological crisis and the threat of nuclear holocaust—must touch all contemporary theologies of history. Both issues, moreover, suggest in graphic terms the more theoretical questions expressed above. For the relationships between science and religion today are not occasioned only by the intriguing and complex set of intellectual issues posed by the new paradigms for method in science and theology and the new content in both disciplines. Nor is the need for cosmology in theology, posed only by the internal theological demand that theology attempt to interpret both self and world in light of its Christic understanding of God; that theology interpret creation even to understand redemption; that theology risk interpreting nature in order to understand history. Central and pressing as these intellectual concerns undoubtedly are, they too must be understood in the context of our contemporary, realistic sense of genuine crisis. The reality of impending ecological crisis is so clear that no serious concern with historical justice can long ignore it. The struggle for justice must also include the struggle for ecology—not only to secure justice for other creatures than the human but even to secure the most basic justice of all: a livable environment for future generations of human beings. The ecological crisis forces all serious political and liberation theology to call into question, on its own praxis criteria, the possible anthropocentrism lurking in its self-understanding. As J. Cobb suggests, all cosmologically-informed theologies (like Cobb's own process theology) must now be in conversation with political and liberation theologians in order to become a political theology. It is also the case that all political and liberation theology must now become an ecological theology as well in order to fulfil its own demands to relate to our present praxis situation. This means that a reopening of cosmological concerns in contemporary theology will also become a dialogue between political theology and those forms of 'post-modern' ecological science (S. Toulmin, F. Ferré) whose own praxis concerns are clear. All must now share a critique and suspicion of traditional scientific and theological understandings of the human right to 'dominate' and exploit nature.

Allied to this sense of concern for the ecological crisis is, of course, the other central praxis issue which our contemporary situation poses: the threat of omnicide present in the possibility of nuclear holocaust. The concern among scientists and theologians alike with this central dilemma imposes, besides its more obvious political implications, a need to interpret cosmology anew. Such interpretations may help to avert this overwhelming and literally final nuclear possibility for the entire planet. The issues—technical, political, and cosmological as well as theological—involved in both ecology and the possibility of nuclear war are so pressing that only a collaborative effort among all persons 'of good will' (including scientists and theologians) can hope to find ways—in theory and praxis—to avert the crisis. For the moment, our point is a more simple and more basic one: viz. that it is now impossible for theology to understand itself as responding to the theoretical and practical challenges of our situation while ignoring the issues of cosmology. If theology is to remain a discipline establishing mutually critical correlations in theory and praxis between interpretations of the situation and interpretations of the tradition, then cosmology must once again be accorded a central place in all theological reflection. Otherwise, we would not be faithful to either the demands of the new intellectual situation in both science and theology nor the crises in praxis which impinge us all. Nor could we claim, under the rubric of a theology of history, that we have adequately interpreted the full resources and demands of the Christian tradition itself; redemption *and* creation; history *and* nature; God, self *and* world. Neither these brief editorial reflections nor the issue as a whole claim to provide resolutions of these issues. What the present *Project X* does pose, however, is a challenge to all theology: a challenge to recognise the new *status quaestionis* which our contemporary situation poses; a challenge to allow, indeed demand that cosmological concerns reenter all contemporary theology. Both the earlier 'warfare' between science and religion and all concordist resolutions are clearly spent. The demand is rather for a collaborative effort that can help establish plausible 'mutually critical correlations' not only to interpret our world but to help change it.

2. NATURE, HISTORY AND HOPE

The dialogue between contemporary political and liberation theologies of history and cosmological theologies in our period could begin with joint reflection on a central category of the Christian theology of history itself: the category of hope. The rediscovery of the prophetic, eschatological and apocalyptic traditions (T. Tshishiku) as bearing political resources could also occasion a retrieval of the cosmic hope present in the wisdom (and cosmological) dimensions of those same traditions. Indeed, in terms of a properly negative function, the theological category of hope challenges both historical ideologies of Western evolutionary liberal progress as well as all cosmologies of progress (whether naturalist-materialist or religious). Both the prophetic and the wisdom traditions recognise what ideologies of progress cannot see: the stark reality of radical evil, tragedy, and sin in all existence and the need for concrete and realistic actions against those evils—action grounded in a fundamental trust—hope in the God of history and nature alike.

There is no possibility in our period for a single, all-embracing 'scientific' cosmological narrative. There is also no possibility, on the theological side, for a complete system of final understanding of God-self-world. What there is, however, is an envisionment of reality informed by the hope afforded by the Christian construal of all reality from the perspective of Jesus Christ. In classical ages where a finite, closed, and seemingly perfectly ordered cosmos seemed plausible, that Christian construal could focus on the incarnation as the always-already immanent and transcendent reality of God's presence to world, self, and history. In our present age all purely incarnational cosmic theologies do not seem a live option. For we have moved from the closed world to an infinite universe (A Koyré). We now recognise more fully the tragic and radical evil in our midst. We fear with reason for the fate of all humanity, indeed of the entire planet in the shadow of massive global suffering and the threats of ecological crisis and nuclear holocaust.

It is not the case, of course, that Christians have ceased to believe in the incarnation of Jesus Christ. It is the case that Christians now recognise that the incarnation itself can only be properly interpreted in the light of the ministry, cross, and resurrection of Jesus Christ. That ministry stands as a stark call to a life of Christian discipleship modelled on the praxis of radically evangelical *imitatio Christi*. The cross intrudes upon all optimism as a stark reminder of the not-yet, the suffering and tragedy at the heart of all existence. The resurrection discloses the reality of Christian hope for nature and history alike as proleptically present in God's vindication of the ministry and cross of Jesus Christ.

It is, therefore, Jesus Christ who continues to provide the decisive clue for a Christian envisionment of God, world, and self, of nature and history, creation and redemption. Yet that theological vision is now recognised to include not only the always-ready reality of the incarnation but also the not-yet realities manifested in the ministry and the cross and the hope for all history and nature disclosed in the resurrection. It is the stark dialect of this always/already/not yet Christic actuality which can inform all Christian theological construals of God, self and world. In so far as Christian theology is the establishment of mutually critical correlations between situation and tradition, this means that theological work must be collaborative with all disciplines competent to interpret both the self (the human sciences including philosophy) and the world (the natural sciences, including philosophy of science and cosmology). In so far as Christian theology is also Christian this means that the incarnation-ministry-cross-resurrection of Jesus Christ will remain the classic Christian clue to an environment of God, self and world in critical correlation with modern theories and contemporary demands of praxis.

The radical theocentrism that should inform all theology as *theo-logos* means, for Christian theology, that self and world will be understood in the light of God. It means as well that the singular, decisive clue to God's own reality and to the realities of self and world (history and nature, redemption and creation) will be understood in relationship to the God disclosed in Jesus Christ. As this Christic reality becomes more and more recognised as the centre of theologies of history and nature alike—as the always/already/not yet reality it is—then the Christian theological category of *hope* will become more and more prominent. For it is theological hope which is likely to become as important in theologies of nature as in theologies of history. Christian hope, focused upon this always/already/not yet reality of Jesus Christ, unmasks historical and cosmic optimism and pessimism alike as unacceptable to Christian faith. The fact is that the easy optimism of earlier concordist cosmologies crashed against the reality of tragedy. And every easy pessimism encourages a quietism which crashes against the reality of genuine Christian hope and the actions demanded by that hope.

Neither optimism nor pessimism but hope is at the heart of the Christian vision of both nature and history. That hope not merely allows but demands both theoretical reflection and concrete praxis. Christian hope grounded in the always/already/not yet reality of Jesus Christ as a primary theological clue for our new cosmological and historical situation. In so far as that situation now demands a theology of history *and* nature, a theology of redemption *and* creation, it suggests that all theologians would do well to focus again on that central category of hope. Then we may begin to see more fully—in critical conversation with modern science, history, philosophy and praxis—some fuller reasons for 'the hope that lies in us' and the beginnings of a new Christian construal of God, self, and world.

All we have attempted to do in these brief reflections is to suggest why the issue of cosmology is both promising and necessary for contemporary Christian theology and some theological resources that may prove helpful when the fuller task begins. That fuller task, we feel confident, receives some new beginnings in the historical and constructive resources presented by the authors of the articles in this issue. To those authors, we express our thanks. To our readers we express our further hope they will join the authors in the kind of reflection needed in our day—a reflection designed to allow the new cosmological concerns to enter the centre of Christian theology and the heart of Christian practice.

Contributors

GÜNTER ALTNER was born in Breslau in 1936. He is a scientist and a theologian: before he became a professor of evangelical theology in Koblenz (1977) he had been professor of human biology in Schwäbisch Gmünd. He is a member of several commissions and institutions concerned with atomic energy, technology, ecology, anthropology and theology. His publications include: *Weltanschauliche Hintergründe der Rassenlehre des Dritten Reiches* (1968); *Die Sonderstellung des Menschen* (1971); *Zwischen Natur und Menschengeschichte* (1975); *Atomenergie—Herausforderung an die Kirchen* (1977); and *Der Darwinismus—Die Geschichte einer Theorie* (1981).

HERMANN BRÜCK. Following early work at the Einstein Institute and Astrophysical Observatory, Potsdam, and a year at the Vatican Observatory, he went to Cambridge in England in 1937 to become assistant director of the Observatory and John Couch Adams Astronomer in the University. In 1947 he moved to Ireland to be senior professor in the Dublin Institute for Advanced Studies and director of the Dunsink Observatory. In 1957 he moved to Scotland as Astronomer Royal for Scotland and regius professor of astronomy in the University of Edinburgh. He retired in 1975 and is now professor emeritus. He has been a member of the Pontifical Academy of Sciences since 1955 and member of the Academy Council since 1964. He has published numerous papers on astrophysics and stellar astronomy in scientific journals and observatory publications. Most recently he has been joint editor with G. V. Coyne and M. S. Longair of *Astrophysical Cosmology. Proceedings of the Study Week on Cosmology and Fundamental Physics* (Pontificia Academia Scientiarum 1982).

JAMES BUCHANAN is lecturer in philosophy and religion at Indiana University N.W., in Gary. He has studied in Moscow, Paris and at Yale University. He is completing his PhD at the University of Chicago in history of religions and the philosophy of religion. His published articles include studies of cross-cultural hermeneutics and East Asian religious thought in comparative perspective.

HENRY CHADWICK is an Anglican priest, now regius professor emeritus at Cambridge University, after having been regius professor of divinity at Oxford between 1959-69 and dean of Christ Church, Oxford, between 1969-79. He is a Fellow of the British Academy; corresponding member of the Académie des Inscriptions et des Belles Lettres; foreign honorary member of the American Academy of Art and Sciences, and of the American Philosophical Society; member of the Anglican/Roman Catholic International Commission 1969-81. His publications include: *Origen contra Celsum* (revised ed. 1980); *The Sentences of Sextus* (1959); *The Early Church* (Penguin 1967); *Early Christian Thought and the Classical Tradition* (1966); *Priscillian of Avila* (1976); *Boethius: the consolations of music, logic, theology, and philosophy* (1981); and *History and Thought of the Early Church* (Variorum 1982).

JOHN COLLINS was born in Ireland in 1946. He gained his PhD from Harvard University in 1972. He is now professor of religious studies at DePaul University, Chicago. His main publications are the following: *The Sibylline Oracles of Egyptian*

Judaism (1974); *The Apocalyptic Vision of the Book of Daniel* (1977); *Daniel, 1 and 2 Maccabees* (1981); *Between Athens and Jerusalem: Jewish Identity in the Hellenistic Diaspora* (1982); and *The Apocalyptic Imagination in Ancient Judaism* (1983).

LANGDON GILKEY was born in 1919. He gained the degrees of AB (Harvard College) (1940), PhD (Columbia University) (1954). Since 1963 he has been professor of theology, Divinity School, University of Chicago. Some of his most recent works are as follows: *Naming the Whirlwind* (1967); *Reaping the Whirlwind* (1976); *Message and Existence* (1980); and *Society and the Sacred* (1981).

MARY HESSE has the degrees of MSc and PhD from the University of London and is a Fellow of the British Academy. She has been professor of the philosophy of science in the University of Cambridge since 1975. She had previously taught in the Universities of London and Leeds, and as visiting professor at Yale, Minnesota, Chicago and Notre Dame. She is a former president of the British Society for the Philosophy of Science, and the Philosophy of Science Association. She was Stanton Lecturer in the University of Cambridge, 1978-80; and will be joint Gifford Lecturer, University of Edinburgh, later this year. Her current interests are metaphor, analogy and symbolism in relation to scientific and religious language; sociology of science and religion. Her principal publications are: *Forces and Fields, a Study of Action at a Distance in the History of Physics* (1961); *Models and Analogies in Physics* (1963); *The Structure of Scientific Interference* (1974); and *Revolutions and Reconstructions in the Philosophy of Science* (1980).

URSULA KING, *née* Brenke, STL (Paris), MA (Delhi), PhD (London), studied theology, philosophy and comparative religion in Germany, France, India and England. From 1965-70 she lived and taught in New Delhi and since 1971 she has been lecturing on Indian religions and sociology of religion in the department of theology and religious studies at the University of Leeds (England). Her research interests concern modern Hinduism, interreligious encounter and the comparative study of mysticism, methodological questions in the study of religion, and women and world religions. Earlier work was in patristics as well as on the thought of Teilhard de Chardin. Besides a number of articles her publications include a German translation of the anti-Arian treatises of Marius Victorinus (undertaken with P. Hadot, Paris), *Christlicher Platonismus* (Zürich 1967) and *Towards a New Mysticism. Teilhard de Chardin and Eastern Religions* (London 1980). Forthcoming is a methodological contribution on 'Historical and Phenomenological Approaches to the Study of Religion. Some Major Developments and Issues under Debate since 1950' in *Contemporary Approaches to the Study of Religion* ed. F. Whaling (Mouton, in press).

NICHOLAS LASH was born in India in 1934. A Roman Catholic, he has been, since 1978, Norris-Hulse professor of divinity in the University of Cambridge. His publications include: *His Presence in the World* (1968); *Change in Focus* (1973); *Newman on Development* (1975); *Voices of Authority* (1976); *Theology on Dover Beach* (1979); and *A Matter of Hope: A Theologian's Reflections on the Thought of Karl Marx* (1981).

OLAF PEDERSEN was born in 1920. He gained a degree in theoretical cosmology in 1943 from the Niels Bohr Institute, Copenhagen, and studied medieval philosophy in Paris with E. Gilson and has since worked in medieval studies and the history of astronomy. From 1956 he has been professor of history of science, Aarhus University, Denmark, and from 1969 visiting fellow of St Edmund's House, Cambridge. He is vice-president of the International Union of the History of Science. Among his publications are *Nicole Oresme* (1956); *Early Physics and Astronomy* (1974); *A Survey of the Almagest* (1975); *Studium Generale* (1979), and minor books and papers.

DAVID TRACY was born in 1939 in Yonkers, New York. He is a priest of the diocese of Bridgeport, Connecticut, and a doctor of theology of the Gregorian University, Rome. He is professor of philosophical theology at the Divinity School of Chicago University. He is the author of *The Achievement of Bernard Lonergan* (1970); *Blessed Rage for Order: New Pluralism in Theology* (1975); and *The Analogical Imagination* (1980). He contributes to several reviews and is editor of the *Journal of Religion* and of the *Religious Studies Review*.

MGR TSHIBANGU TSHISHIKU is auxiliary bishop of Kinshasa, Zaïre. Born near Lubumbashi in 1933, Mgr Tshishiku was educated in Zaïre and in Louvain, where he gained a licenciate and two doctorates in theology. His academic interests are theological methodology, the philosophy of science and comparative religion, and he has written widely on all these topics, and especially on the African contribution to Christianity. He is a member of the Roman secretariat for non-Christian religions, a former member of the Faith and Order Commission of the World Council of Churches, and head of the episcopal theological commissions of Zaïre and all Africa. Mgr Tshishiku has participated in the administration of numerous African universities, is a consultant for UNESCO, and sits on the editorial committee for the *General History of Africa*. He is a Commander of Zaïre's National Order of the Leopard and of the National Order of Senegal.

WILLIAM G. WARTHLING was born in Buffalo, New York, in 1936. He studied at Niagara University and was ordained priest for the diocese of Buffalo. He holds the STB and STL degrees from the Gregorian University in Rome. Presently he teaches philosophy of religion at Niagara County Community College and at the Attica State Correctional Facility. He has been a Board of Trustee member at the Lonergan Institute since its inception and is an appointee of the Governor of New York State as an overseer of mental health care in Buffalo. He is a well-known lecturer on the relationships of art, religion and science at many American institutions.

CONCILIUM

All back issues are still in print and available for sale. Orders should be sent to the publishers,

T. & T. CLARK LIMITED
36 George Street, Edinburgh EH2 2LQ, Scotland

GOD IS NEW EACH MOMENT

Edward Schillebeeckx

IN CONVERSATION WITH
HUUB OOSTERHUIS & PIET HOOGEVEEN

In response to the probing questions of his colleagues, Edward Schillebeeckx provides a fascinating and comprehensible overview of his intellectual development and the concrete implications of the major themes in his work. **GOD IS NEW EACH MOMENT** permits an encounter with the flesh-and-blood Schillebeeckx—a man whose thinking is driven by his passionate concern to live a gospel Christianity that is engaged with the great social, political, and intellectual issues of the modern world. Clearly distilled are his ideas about Jesus, the Scriptures, ministry and sacraments, the future of the Church, the feminist movement, the liberation of the poor. **GOD IS NEW EACH MOMENT** explores the sources of Schillebeeckx' thought: the people, ideas, and experiences that have shaped his work.

144 pages   published in paperback

in the United States & Canada   in the United Kingdom
꒐SEABURY PRESS   T. & T. Clark, Ltd.
Seabury Service Center · Somers, CT 06071

THE
GLORY
OF THE
LORD

A Theological Aesthetics

HANS URS VON BALTHASAR

Probably the most important sustained piece of theological writing to appear since Karl Barth's *Church Dogmatics,* von Balthasar's work restores aesthetics and contemplation to their rightful place in Christian theology.

Hans Urs von Balthasar is one of the magisterial figures of contemporary theology and posseses a remarkable knowledge of the theological and metaphysical traditions as well as of Western letters. In the 7 volumes of *The Glory of the Lord* he shows how the Biblical vision of the divine glory, revealed in the crucified and risen Christ and reflected in the great theologians of the Christian tradition, fulfils and transcends the perception of Being in Western metaphysics.

In **Seeing The Form**, the first of the 7 volumes, von Balthasar reviews the developments in Protestant and Catholic theology since the Reformation, then turns to the central theme of the volume—the question of theological knowledge.

Vol. 1 is available now   656pp   cased   £19·95

(Published in U.S.A. by Ignatius Press/Crossroad Publishing)

T & T CLARK LTD, 36 GEORGE STREET, EDINBURGH EH2 2LQ, SCOTLAND